MAPPING THE NATION

SUPPORTING DECISIONS THAT GOVERN A PEOPLE

Esri Press, 380 New York Street, Redlands, California 92373-8100

Ask for Esri Press titles at your local bookstore or order by calling 800-447-9778, or shop online at esri.com/esripress. Outside the United States, contact your local Esri distributor or shop online at eurospanbookstore.com/esri.

Esri Press titles are distributed to the trade by the following:

In North America:
Ingram Publisher Services
Toll-free telephone: 800-648-3104
Toll-free fax: 800-838-1149
E-mail: customerservice@ingrampublisherservices.com

In the United Kingdom, Europe, Middle East and Africa, Asia, and Australia:
Eurospan Group   Telephone: 44(0) 1767 604972
3 Henrietta Street Fax: 44(0) 1767 601640
London WC2E 8LU   E-mail: eurospan@turpin-distribution.com
United Kingdom

CONTENTS

CONTENTS

FOREWORD

National events can range from great celebrations, such as the president's inauguration, to devastating catastrophes like Hurricane Sandy along the Atlantic Coast. These events underscore the vital role geographic information systems (GIS) play in supporting decisions, securing public safety, and solving problems.

At the presidential inauguration, for example, a GIS-based enterprise platform integrated with numerous information systems helped to provide security for the 2.7 million people who attended. The hallmark of geospatial technology has always been its unique ability to bring information together collaboratively. The cloud-based platform we now have in place expands that mission with efficiency, accuracy, and a pervasiveness that was unfathomable just a few years ago.

The USDA Forest Service and other federal agencies pioneered the use of GIS decades ago, and these organizations have kept pace with a technology that has evolved to include real-time, dynamic imagery and 3D capabilities that significantly enhance spatial analyses. Today's GIS platform represents a huge shift to online services that have enhanced open government and vastly broadened the sharing of maps and data across all branches and levels of government.

This volume of *Mapping the Nation*, our annual federal map book, is filled with excellent maps and applications that vividly convey the power and accessibility of GIS.

Federal agencies have long been the proving ground for applying GIS to a wide variety of workflows and problems. In these pages, you will see how great cartography can help tell compelling stories, how GIS is advancing science across multiple disciplines, how the government relies on GIS for planning and managing, and how this era of big data presents limitless opportunities to transform our nation and our world.

These are only a few of the many great examples where the federal government was able to leverage technology and geographic knowledge to better serve the country and its citizens. I hope you enjoy it.

Warm regards,

Jack Dangermond,
President, Esri

"We're on the cusp of a new golden age when it comes to harnessing the power of data—government data—to make a positive impact on our nation.... Rich, powerful, interactive maps are becoming truly ubiquitous tools."

TODD PARK
Chief Technology Officer for the United States of America

INTRODUCTION

What is GIS?

A geographic information system (GIS) is location-based technology for organizing and integrating data into meaningful information that answers questions, predicts outcomes, and solves problems. This information is communicated through analytical maps and imagery increasingly accessible anytime and anywhere. Federal agencies have long relied on GIS to analyze complex situations, visualize problems, and develop geographic strategies. Redundancies, inefficiencies, outdated or inaccessible data, poor communication, and limited collaboration all serve to increase costs and diminish performance. GIS helps agencies eliminate those institutional shortcomings while meeting the growing demand for open government and accountability to citizens.

Return on investment: The ultimate benefit

The US federal government has made major investments in geospatial data and technology, knowing that the nation's policies, actions, and events are best understood within a geographic context. *Mapping the Nation: Supporting Decisions that Govern a People* shows many examples of how GIS has evolved into a full-fledged platform so that data, resources, and services are now easily accessed, leveraged, and shared within and among federal agencies.

Government organizations realize a significant return on their GIS investments. GIS demonstrates real business value in such diverse fields as environmental management, health care, education, law enforcement, emergency response, national defense, and transportation. GIS is integral to operations, decision making, and communication, resulting in long-term savings, streamlined workflows, enhanced collaboration, and improved overall efficiency. The versatility, agility, and flexibility of GIS, particularly in rapidly evolving web and mobile applications, helps federal organizations face complex challenges while serving the nation's citizens. One of the most significant examples of return on investment is that of the Federal Aviation Administration.

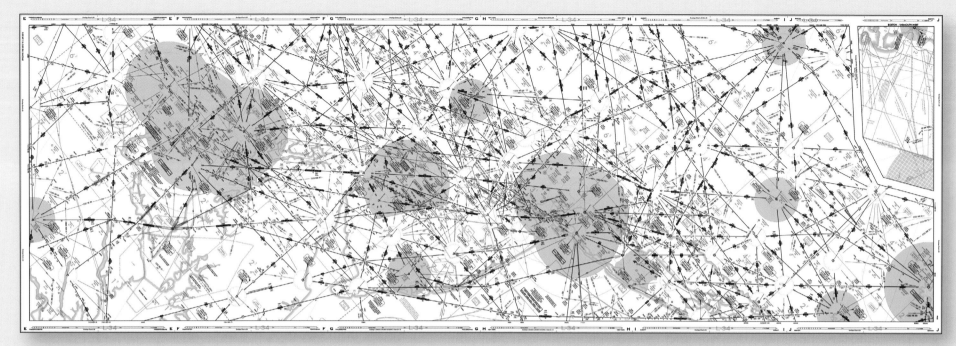

FAA Aeronav Products: Promoting Safety While Reducing Costs

To address high labor costs while maintaining high-quality products, AeroNav and the FAA sought to standardize the production of flight charts. Investment in GIS has been driving efficiency gains of these map products.

AeroNav Products leverage their investment in GIS by running parallel efforts throughout the FAA. This unified approach will enable the organization to keep the program viable, both financially and technologically.

PROJECT BENEFITS

AeroNav predicts that once the improved charting is implemented, a 41 percent labor savings will be realized, while also improving accuracy and product quality.

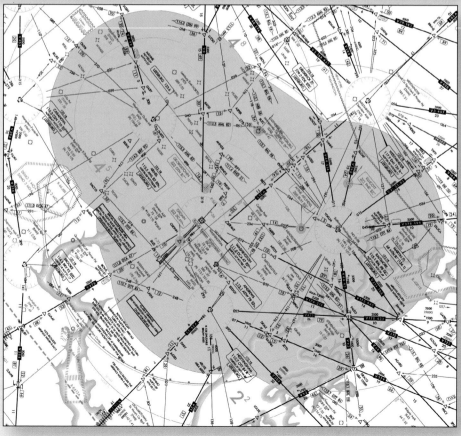

Low altitude chart showing navigation features below 18,000 feet in the Washington, DC, metropolitan area.

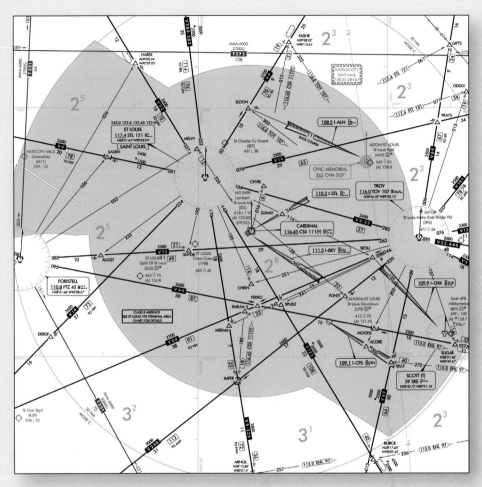

Close-up of a low altitude chart designed to show congested areas at a larger scale in St. Louis, Missouri.

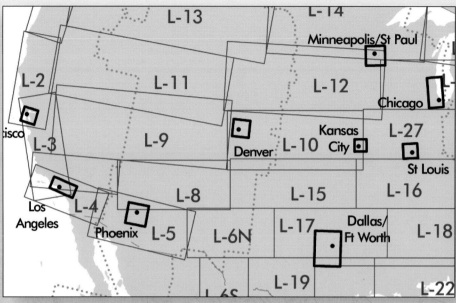

Index of low altitude area charts.

Learn more about the maps

QR code readers are available at any app store. Scan the codes thoughout the book to directly access related websites and learn more about the maps and organizations behind them. Some content may require Adobe Flash Player.

http://www.faa.gov/air_traffic/flight_info/aeronav/

US DEPARTMENT OF AGRICULTURE

"Putting our data on these maps within our existing systems is a really, really big deal for us.... Just clicking on a map is a more intuitive way for us to understand and to explore our own data."

TODD SCHROEDER
Director of Business Systems Management of the US Department of Agriculture
Animal and Plant Health Inspection Service

GIS Support Maps for National Level Exercise and National Radiation Exercise

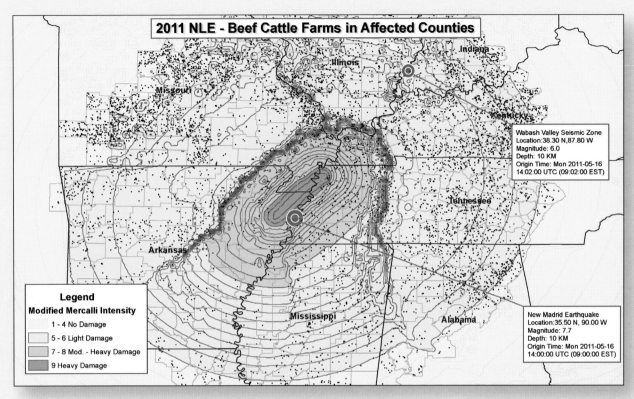

We often think of disasters and their effects on cities, but planners must also consider the effects on the nation's farms. For the National Level Exercise 2011, ArcGIS maps show affected beef cattle farms and poultry and egg production farms that tested state and federal response to a magnitude 7.7 earthquake. These maps prepared with GIS help various agencies plan and prepare for disasters.

http://www.dm.usda.gov/ohsec/

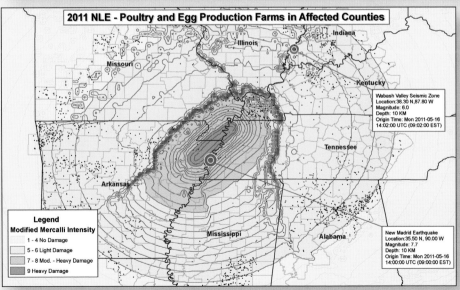

US National Arboretum Botanical Explorer

PROJECT BENEFITS

This application provides the National Arboretum staff with intuitive, one-stop access to data, greatly expediting research. A kiosk in the visitor center helps visitors navigate the grounds, saving hours of staff time answering questions.

Scientists, horticulturists, researchers, and the public want interactive maps of the National Arboretum. The Arboretum Botanical Explorer links plant records and images in a user-friendly display. Web-based ArcGIS drives the mapping functions that provide access to plant data, major gardens and collections, and unique features such as sculptures and structures. Researchers use the site to access every plant and associated data, and visitors use it for services and self-guided tours. The flexibility of GIS permits a variety of uses to benefit all users of the National Arboretum.

http://www.usna.usda.gov/

Coconino National Forest Initial Attack Map Book

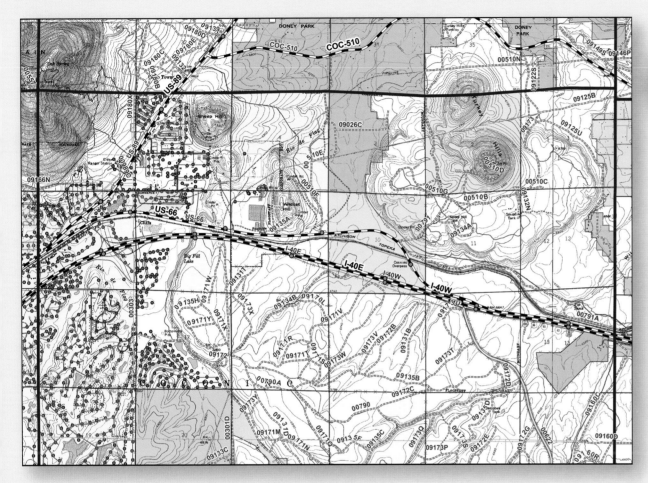

PROJECT BENEFITS

Fire management on the Coconino National Forest now has a comprehensive, detailed, and easy-to-navigate map set that enables cost-effective fire suppression and resource-management efforts.

The Forest Service has to know how to react quickly and plan efficiently when fires break out, so it created this set of maps. Available in print, on mobile devices, and the dispatch website, these easy-to-access maps help staff improve emergency response plans and reduce forest fire damage.

http://esriurl.com/7198

Golden-Winged Warbler Habitat Modeling

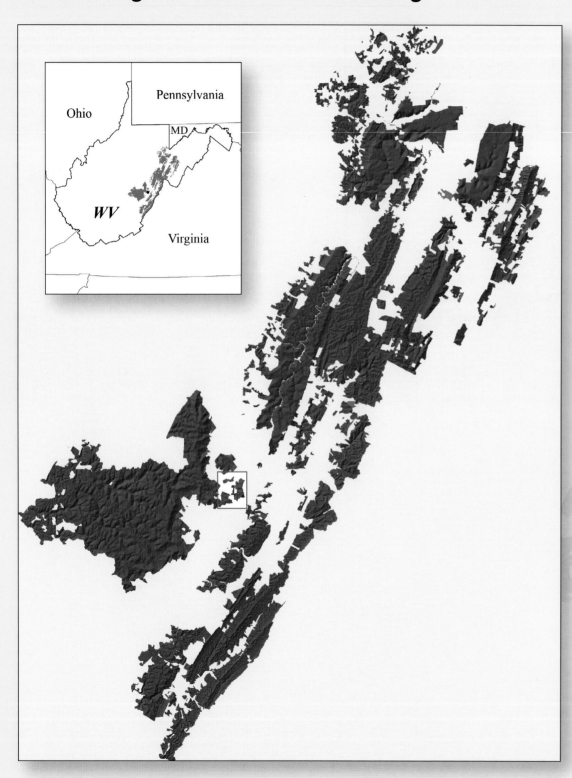

The golden-winged warbler is a critically threatened bird species, and conservation efforts are under way to protect them. Developing a thorough understanding of their habitat structure may be the best way to construct a plan to manage a habitat for the remaining populations. Modeling their habitat using satellite imagery and GIS, as shown here, may be a crucial step toward saving them from becoming endangered.

http://gwwa.org/wvu.html

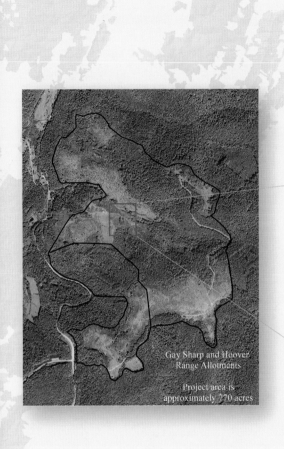

Gay Sharp and Hoover
Range Allotments

Project area is
approximately 770 acres

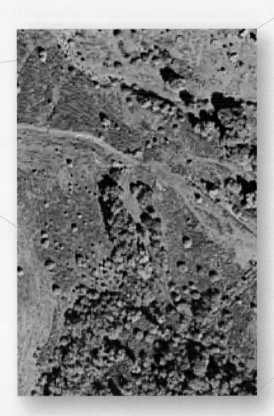

Profile

Great Lakes Conservation Region
Appalachian Conservation Region
Golden-winged Warbler predicted occurrence
Blue-winged Warbler predicted occurrence
Golden-wing/Blue-wing predicted overlap

0 125 250   500

Profile View

Elevation: Start = 967.157; End = 997.908; Difference = 30.751 <Meter>

Determining height

LAS Dataset P R O F I L E

Best Management Opportunity Model—Restoring Chestnut to the Landscape

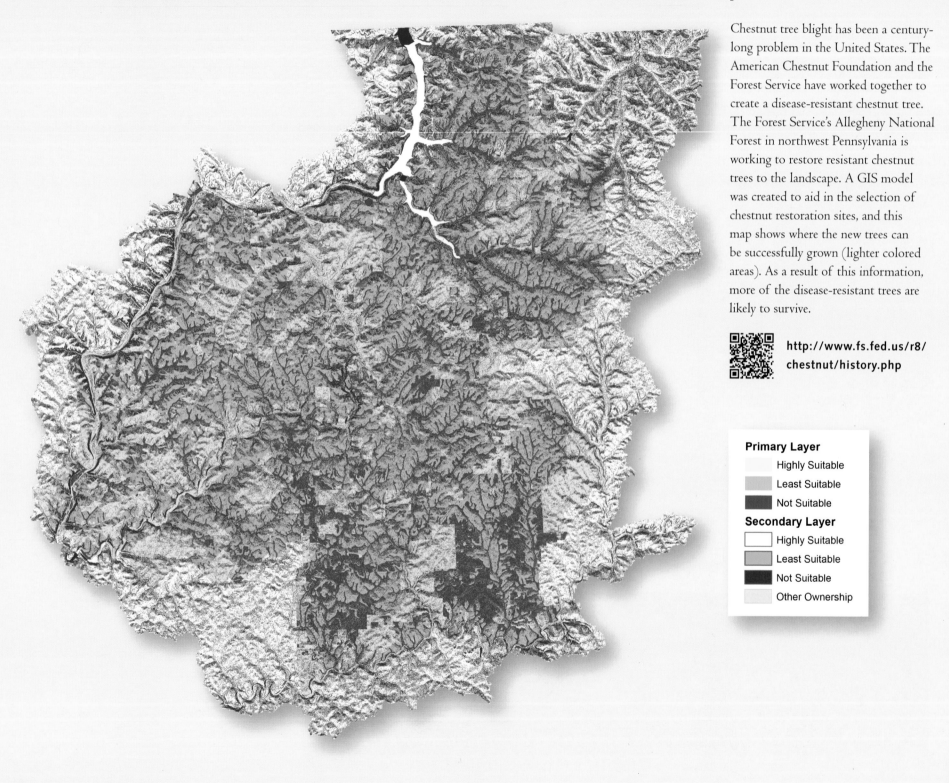

Chestnut tree blight has been a century-long problem in the United States. The American Chestnut Foundation and the Forest Service have worked together to create a disease-resistant chestnut tree. The Forest Service's Allegheny National Forest in northwest Pennsylvania is working to restore resistant chestnut trees to the landscape. A GIS model was created to aid in the selection of chestnut restoration sites, and this map shows where the new trees can be successfully grown (lighter colored areas). As a result of this information, more of the disease-resistant trees are likely to survive.

http://www.fs.fed.us/r8/
chestnut/history.php

Primary Layer
- Highly Suitable
- Least Suitable
- Not Suitable

Secondary Layer
- Highly Suitable
- Least Suitable
- Not Suitable
- Other Ownership

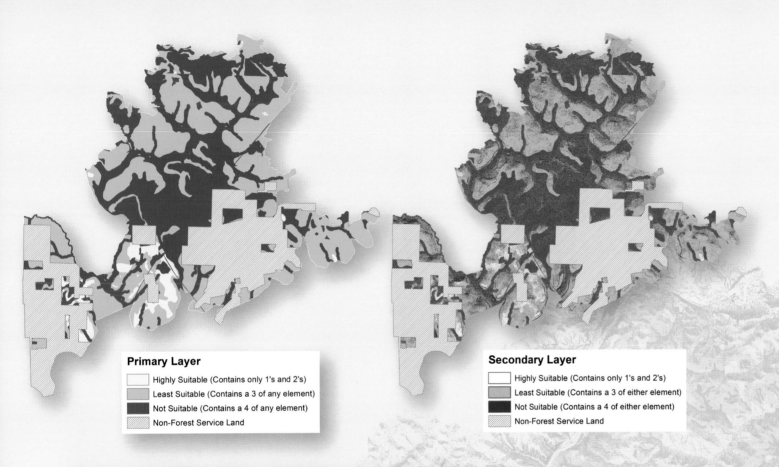

Primary Layer

- Highly Suitable (Contains only 1's and 2's)
- Least Suitable (Contains a 3 of any element)
- Not Suitable (Contains a 4 of any element)
- Non-Forest Service Land

Secondary Layer

- Highly Suitable (Contains only 1's and 2's)
- Least Suitable (Contains a 3 of either element)
- Not Suitable (Contains a 4 of either element)
- Non-Forest Service Land

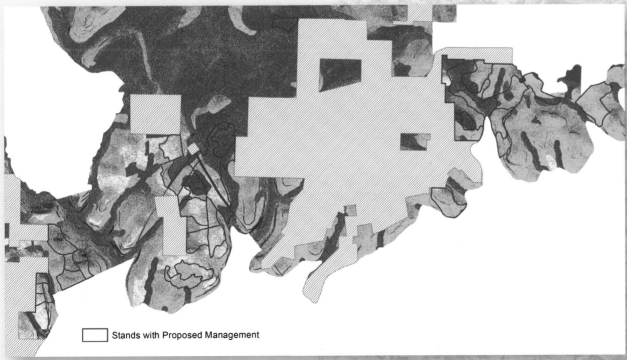

- Stands with Proposed Management

Wallow Fire: Bringing Partners and Data Together

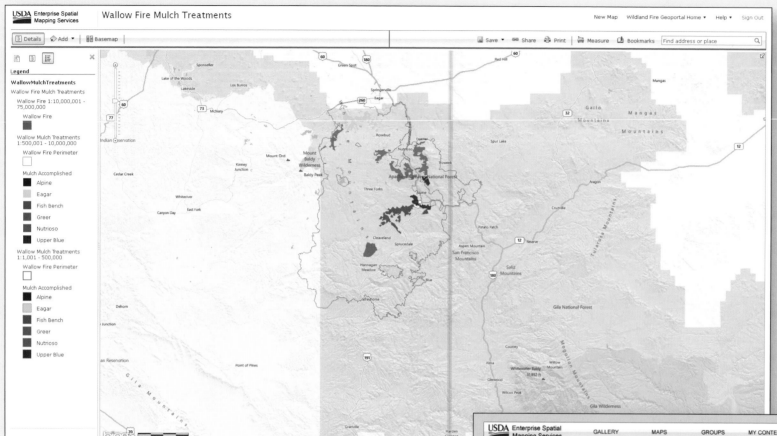

The Forest Service needed a way to help agencies work together to manage fire recovery. To facilitate collaboration, the Forest Service used a GIS to build the Wildland Fire Geoportal, a website to share data and information about the 2011 Wallow Fire in northeast Arizona. This collection of online maps allows agencies to view, share, and analyze data from a single web and spatial interface.

http://www.fs.usda.gov/asnf

Vegetation Height Santa Fe National Forest

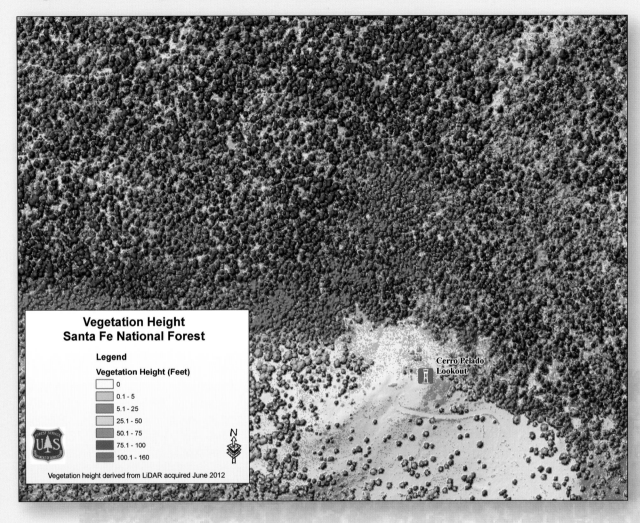

**Vegetation Height
Santa Fe National Forest**

Legend

Vegetation Height (Feet)

- 0
- 0.1 - 5
- 5.1 - 25
- 25.1 - 50
- 50.1 - 75
- 75.1 - 100
- 100.1 - 160

Vegetation height derived from LiDAR acquired June 2012

Cerro Pelado
Lookout

A large-scale, long-term project to restore sustainable ecological forest conditions is under way in the Southwest Jemez Mountains. The Forest Service used GIS and lidar to identify areas of dense stands of trees in need of thinning or prescribed fire to reduce fuel loads and help reduce the chance of catastrophic wildfire. The vegetation height data derived from lidar was also used to find areas with taller trees that were likely old growth stands. By identifying ecological components such as old growth and dense stands of trees, project planners are able to more successfully strategize forest restoration efforts.

 http://esriurl.com/7199

Maah Daah Hey National Trail Map

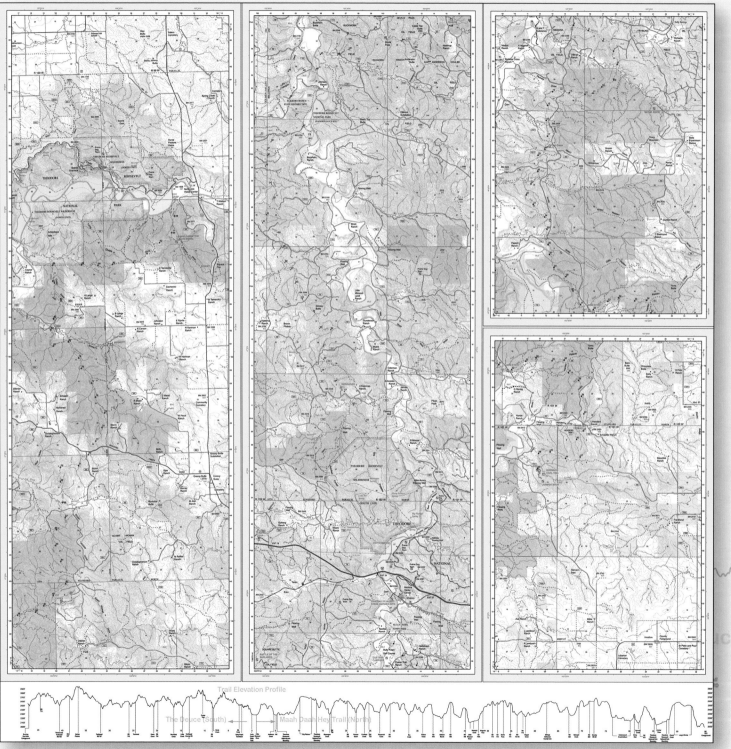

The Forest Service promotes sustainable community-based tourism on National Forest lands by providing activities such as bird watching, hiking, and cycling. To attract visitors, the Forest Service created an easy-to-use and detailed trail map for the Little Missouri National Grasslands of North Dakota. GIS is used here to develop a product formatted and designed especially for hikers. Targeted guides like this help attract tourism and boost local economies by publicizing appealing activities to visitors of the area.

http://esriurl.com/7200

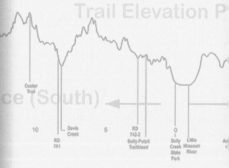

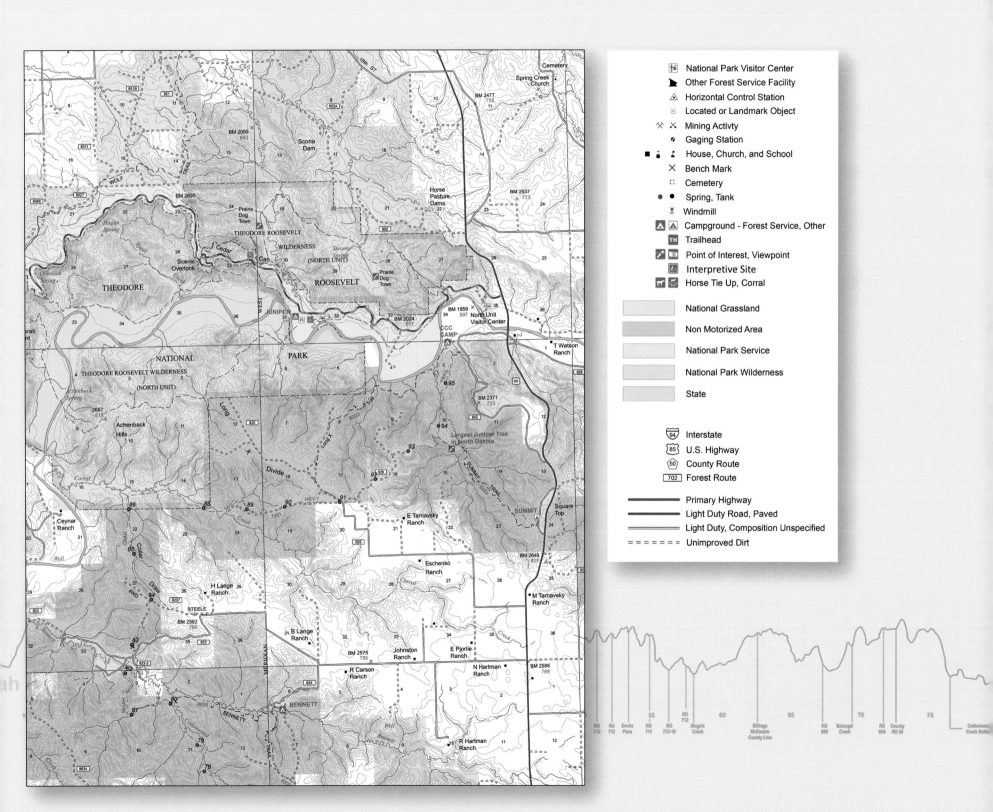

National Park Visitor Center
Other Forest Service Facility
Horizontal Control Station
Located or Landmark Object
Mining Activty
Gaging Station
House, Church, and School
Bench Mark
Cemetery
Spring, Tank
Windmill
Campground - Forest Service, Other
Trailhead
Point of Interest, Viewpoint
Interpretive Site
Horse Tie Up, Corral

National Grassland
Non Motorized Area
National Park Service
National Park Wilderness
State

Interstate
U.S. Highway
County Route
Forest Route

Primary Highway
Light Duty Road, Paved
Light Duty, Composition Unspecified
Unimproved Dirt

Forest Pest Conditions

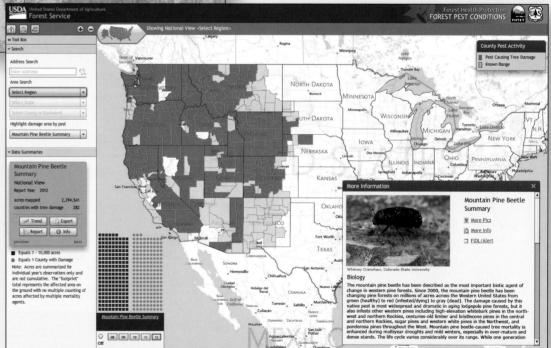

PROJECT BENEFITS

GIS technology allows for the Forest Service to develop and deliver authoritative information gathered nationally in an efficient and cost-effective manner.

The Forest Service needed a way to manage and distribute pest information that covers an enormous area and is used by many people. Using a GIS, the Forest Service developed pest damage databases and keeps them up to date. This website shows county-level maps of major forest insect and disease conditions throughout the United States. The ability to compile and publish timely information plays a significant role in managing national resources.

http://foresthealth.fs.usda.gov/portal

Decision Support for Aerial Fire Retardant Avoidance

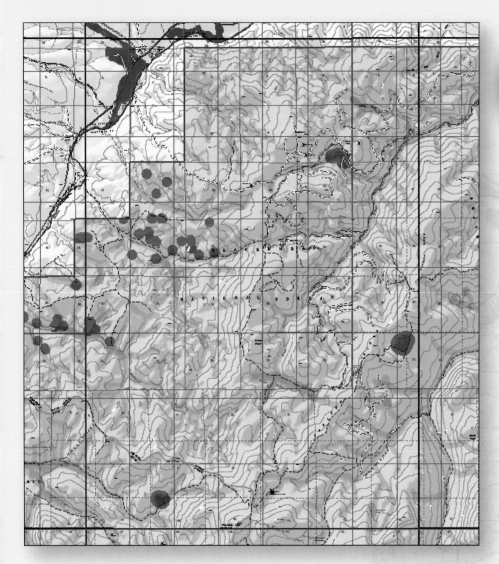

PROJECT BENEFITS

Using GIS to standardize data and processes streamlines the production of maps and improves access to resources for the fire community, both internal and external to the Forest Service.

The Forest Service needs to provide comprehensive, up-to-date information for aerial fire response teams. To accomplish that goal, GIS was used to automatically generate nearly 11,000 regional maps that cover all national forests. This image shows one of those regional maps, the Foolhen Mountain area within the Beaverhead-Deerlodge National Forest in southwest Montana.

http://www.fs.fed.us/fire/retardant/index.html

2010 Monitoring Trends in Burn Severity for Swakane, WA

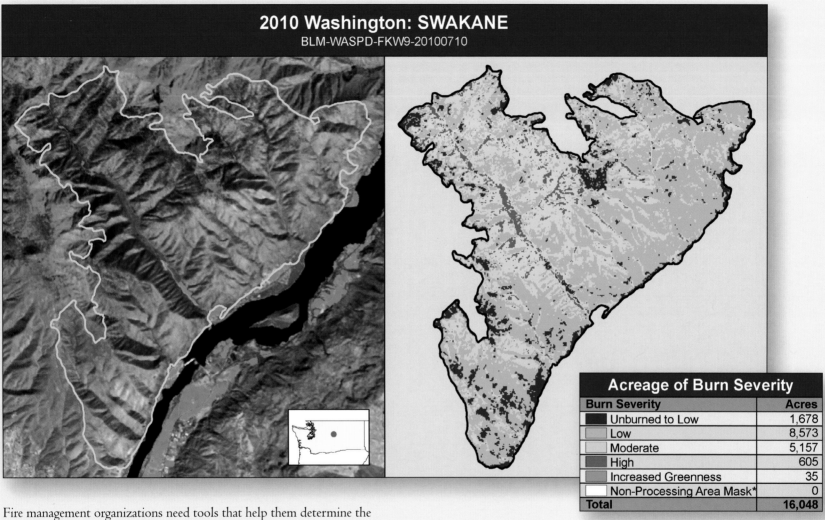

2010 Washington: SWAKANE
BLM-WASPD-FKW9-20100710

Acreage of Burn Severity	
Burn Severity	**Acres**
Unburned to Low	1,678
Low	8,573
Moderate	5,157
High	605
Increased Greenness	35
Non-Processing Area Mask*	0
Total	**16,048**

Fire management organizations need tools that help them determine the severity and scope of wildfire damage as quickly as possible. As shown here, GIS can be combined with post-fire satellite imagery to show how badly vegetation was burned and the amount of land affected. Such images contribute significantly to effective disaster management and response.

http://activefiremaps.fs.fed.us/gisdata.php

Fire Detections and Forest Service Lands: 2001 to 2011

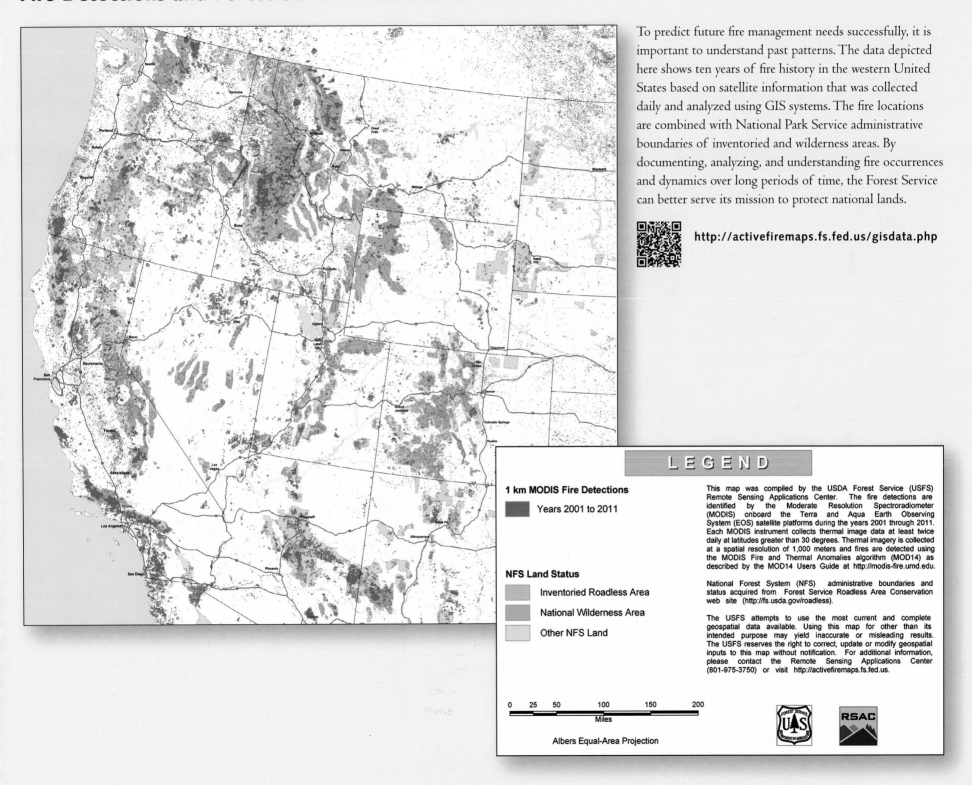

To predict future fire management needs successfully, it is important to understand past patterns. The data depicted here shows ten years of fire history in the western United States based on satellite information that was collected daily and analyzed using GIS systems. The fire locations are combined with National Park Service administrative boundaries of inventoried and wilderness areas. By documenting, analyzing, and understanding fire occurrences and dynamics over long periods of time, the Forest Service can better serve its mission to protect national lands.

http://activefiremaps.fs.fed.us/gisdata.php

LEGEND

1 km MODIS Fire Detections

Years 2001 to 2011

NFS Land Status

Inventoried Roadless Area

National Wilderness Area

Other NFS Land

This map was compiled by the USDA Forest Service (USFS) Remote Sensing Applications Center. The fire detections are identified by the Moderate Resolution Spectroradiometer (MODIS) onboard the Terra and Aqua Earth Observing System (EOS) satellite platforms during the years 2001 through 2011. Each MODIS instrument collects thermal image data at least twice daily at latitudes greater than 30 degrees. Thermal imagery is collected at a spatial resolution of 1,000 meters and fires are detected using the MODIS Fire and Thermal Anomalies algorithm (MOD14) as described by the MOD14 Users Guide at http://modis-fire.umd.edu.

National Forest System (NFS) administrative boundaries and status acquired from Forest Service Roadless Area Conservation web site (http://fs.usda.gov/roadless).

The USFS attempts to use the most current and complete geospatial data available. Using this map for other than its intended purpose may yield inaccurate or misleading results. The USFS reserves the right to correct, update or modify geospatial inputs to this map without notification. For additional information, please contact the Remote Sensing Applications Center (801-975-3750) or visit http://activefiremaps.fs.fed.us.

0   25   50   100   150   200
Miles

Albers Equal-Area Projection

Rapid Assessment of Vegetation Condition after Wildfire

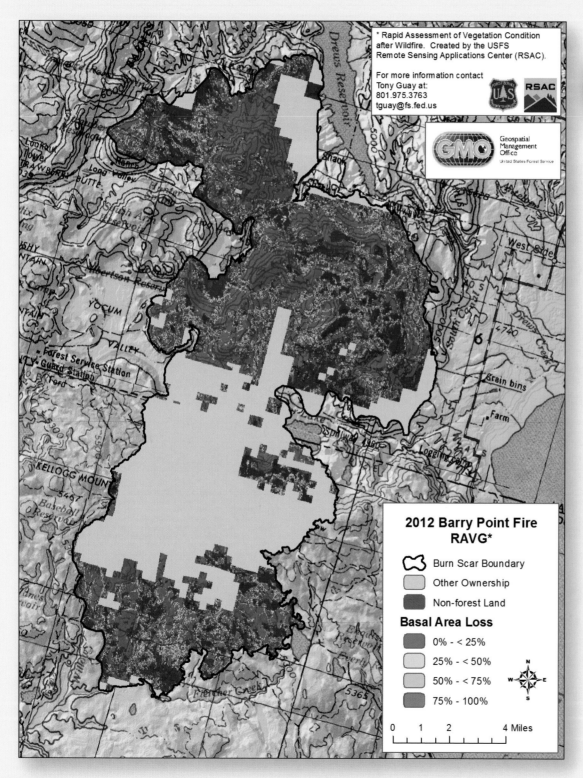

* Rapid Assessment of Vegetation Condition after Wildfire. Created by the USFS Remote Sensing Applications Center (RSAC).

For more information contact
Tony Guay at:
801.975.3763
tguay@fs.fed.us

Geospatial
Management
Office
United States Forest Service

**2012 Barry Point Fire
RAVG***

- Burn Scar Boundary
- Other Ownership
- Non-forest Land

Basal Area Loss

- 0% - < 25%
- 25% - < 50%
- 50% - < 75%
- 75% - 100%

0   1   2   4 Miles

PROJECT BENEFITS

Investing in remote sensing and GIS technology improves the management of the National Forest System lands with more efficient and fiscally responsible decisions following a wildfire.

To facilitate forest vegetation restoration after a wildfire, it is crucial to accurately identify post-fire vegetation severity. By identifying the areas where forest vegetation was affected after the 2012 Barry Point Fire in south central Oregon, this map uses GIS to identify areas of concern using a satellite image-based method of mapping vegetation conditions. This helps ensure cost-effective and timely reforestation and restoration efforts can be implemented within burned areas.

http://www.fs.fed.us/eng/rsac/

Lidar Applications in the USDA Forest Service

In the past, scientists and technicians spent lots of time and money performing physically demanding and often dangerous field work to gather data about waterways, wildlife habitats, and forest inventory. These images showcase recent examples of how that time-consuming field work has been made more efficient by using GIS and airborne lidar data.

http://www.fs.fed.us/eng/rsac/

Lidar Point Cloud

Hydrology

Bare Earth DEM

Synthetic Drainage Network

Using ArcHydro extension in ArcGIS to derive synthetic drainage networks from 1 meter lidar derived DEMs.

Habitat Modeling

Canopy Height

Canopy Cover

Habitat Model

Red Squirrel Habitat (red pixels) was defined in Spatial Analyst as areas that exceed 70 feet in Canopy Height and greater than 50% Canopy Closure.

Forest Structure

Field Plot Measurements

Lidar Plot Statistics

Statistical Modeling

Continuous Basal Area

Live Basal Area was calculated over the entire lidar project area using Spatial Analyst to apply the modeled relationship between the field plot data and the lidar plot statistics.

23

Potential Areas Suitable for Forest Restoration Projects

PROJECT BENEFITS

Automated and refined data processing tasks help decision makers evaluate where investments in restoration would yield the greatest benefit.

The Geospatial Service and Training Center uses GIS to identify areas suitable for forest restoration in the western United States. This map has the potential to improve ecosystem health and contribute to local economies because more informed decisions about return on investment can be made.

http://www.fs.fed.us/gstc/

Forest Service Boundaries
- Potential Restoration Areas
- National Forest System Lands
- Administrative Regions

Other Political Boundaries
- States

The data shown in this map are derived from coarse scale data and may not reflect actual suitablity in all cases.

Prepared By: Greg Hughes, Ted Kwasnik   May 25, 2013

The US Forest Service Interactive Visitor Map

PROJECT BENEFITS

It is important that visitors are provided with maps and applications that are accurate, accessible, and easy to navigate. The Forest Service recognizes that comprehensive web-based maps are the most efficient and cost-effective way to provide the public with information needed to experience national forests and grasslands in a safe manner.

The Forest Service wants to make sure that visitors have the information they need to easily find the sites they would like to explore. To ensure a positive visitor experience, the Forest Service used GIS to create a simple, intuitive web interface for guest use. The Interactive Visitor Map allows guests to search for and locate their areas of interest. Once a sightseer finds a destination, he or she can view detailed information, create maps, and find the fastest route to get there.

http://foresthealth.fs.usda.gov/portal

USDA NASS Cropland Data Layer

Accurate information on crop-specific land cover is necessary to estimate acreage for agricultural lands. GIS is used to automate the processing that accurately and objectively measures the percentage of crop cultivation in the United States, and the CropScape portal is how the data is accessed and distributed. The automated process made possible with GIS improves access to the information.

http://nassgeodata.gmu.edu/CropScape

US DEPARTMENT OF COMMERCE

"(Storm surge mapping is) really about communication... It's helping the decision makers envision what will be the impacts on the ground and what type of impacts they would expect from a hurricane."

JAMIE RHOME
Storm Surge Unit Team Lead, National Hurricane Center

Sea Level Rise and Coastal Flooding Impacts Viewer

Many communities need to understand the impact that coastal flooding can have in order to properly plan and prepare. This map viewer of Galveston, Texas, allows users to see how and where sea level rise and coastal flooding will occur. Seeing the potential effects on cities and towns can have a big impact on decisions that help to save lives and property.

http://www.csc.noaa.gov/digitalcoast/ tools/slrviewer

Offshore Wind Development Potential

Planners need a visual representation of the potential development of offshore wind energy. This online map shows energy and infrastructure data along the eastern seaboard. Using an online platform to host the map data allows decision makers from multiple agencies to quickly access information.

  http://marinecadastre.gov/

Commercial Vessel Density, October 2009–2010

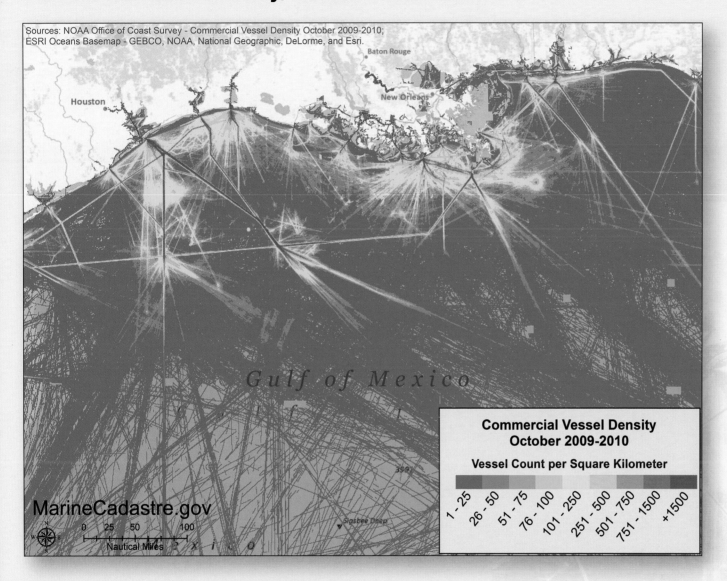

Sources: NOAA Office of Coast Survey - Commercial Vessel Density October 2009-2010;
ESRI Oceans Basemap - GEBCO, NOAA, National Geographic, DeLorme, and Esri.

Baton Rouge

Houston

New Orleans

Gulf of Mexico

MarineCadastre.gov

0   25   50   100
Nautical Miles

**Commercial Vessel Density
October 2009-2010**

Vessel Count per Square Kilometer

1 - 25 26 - 50 51 - 75 76 - 100 101 - 250 251 - 500 501 - 750 751 - 1500 +1500

The ability to repurpose map information was important when examining commercial vessel traffic patterns in the Gulf of Mexico from October 2009 to October 2010. The US Coast Guard and National Oceanic and Atmospheric Administration collaborated to create a web map service that provides improved accessibility and performance. Set up as a service, map information can be used in other applications, websites, and publications to improve efficiency and benefit more users.

 http://marinecadastre.gov/

33

NOAA Gulf of Mexico Data Atlas

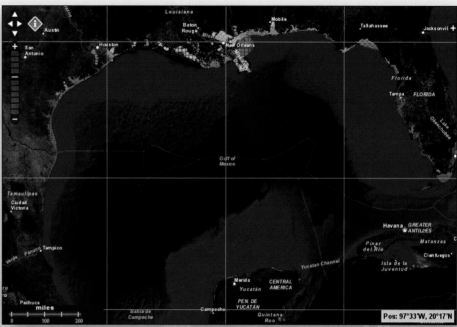

Access to GIS mapping data is crucial to ecosystem modeling, habitat assessment, and monitoring efforts undertaken by the National Oceanic and Atmospheric Administration and its partners. Maps displayed in the Gulf of Mexico Data Atlas show the long-term status and trends of Gulf ecosystems and enable users to visualize complex datasets and create their own map products. This represents a collaborative effort of more than thirty federal, Gulf state, and local agencies, along with academia, nongovernmental organizations, and international partners. The online delivery of maps and data gives agencies easy access to critical information to manage the Gulf of Mexico.

http://gulfatlas.noaa.gov/

Electronic Navigational Charts Portal

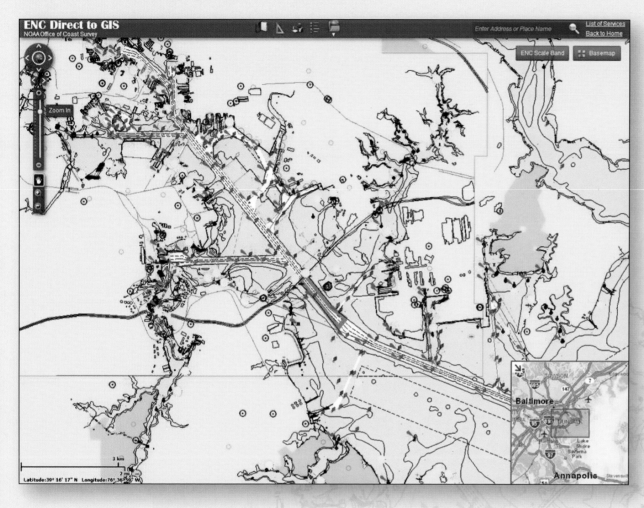

Up-to-date navigational charts are essential to the safe use and management of the nation's coastal waterways. This web-based charts portal provides public access to the entire suite of data and charts that are powered by GIS mapping. From harbor cruise to harbor management, current and clear charts that readily inform people help in the proper use of the nation's waterways.

http://www.nauticalcharts.noaa.gov/ csdl/ctp/encdirect_new.htm

Integrated Ocean and Coastal Mapping Sandy Coordination

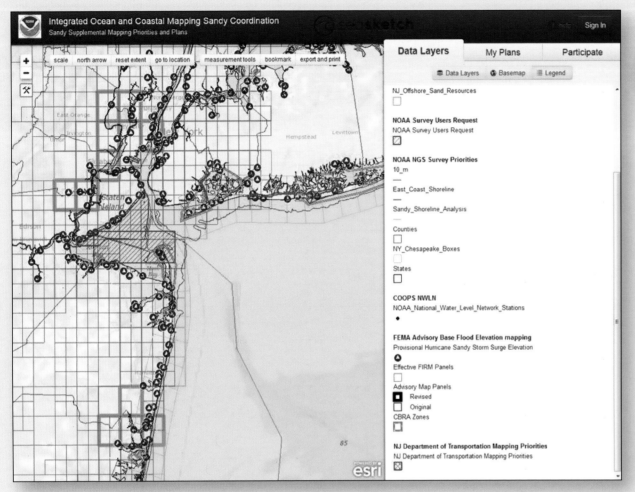

PROJECT BENEFITS

This GIS application helped to ensure that federal emergency funds were directed to those areas impacted by the storm. SeaSketch also averted redundant mapping efforts for a more coordinated and integrated ocean and coastal mapping approach.

Natural disasters make the quick and easy collection of needs and recovery plans imperative. Mapping data is often essential to assess damage and aid recovery. Using the SeaSketch collaboration tool, the National Oceanic and Atmospheric Administration and its agency partners were able to quickly gather data needs and mapping plans from federal and state agencies for areas impacted by Hurricane Sandy.

http://oceanservice.noaa.gov/news/weeklynews/oct12/
nos-response-sandy.html

Net Migration Flows for Dane County, Wisconsin

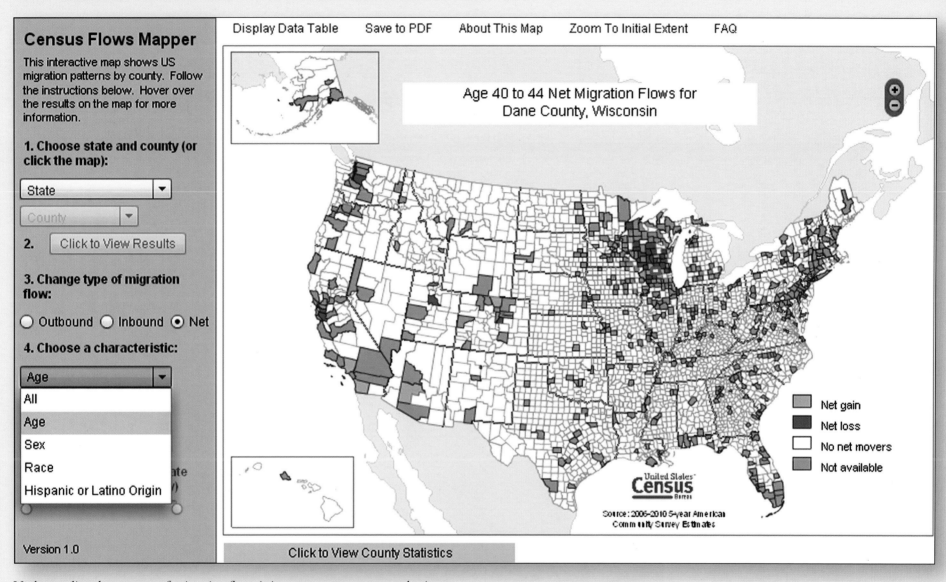

Census Flows Mapper

This interactive map shows US migration patterns by county. Follow the instructions below. Hover over the results on the map for more information.

1. Choose state and county (or click the map):

State ▼

County ▼

2. Click to View Results

3. Change type of migration flow:

○ Outbound ○ Inbound ● Net

4. Choose a characteristic:

Age ▼

All

Age

Sex

Race

Hispanic or Latino Origin

Version 1.0

Display Data Table   Save to PDF   About This Map   Zoom To Initial Extent   FAQ

Age 40 to 44 Net Migration Flows for Dane County, Wisconsin

Net gain

Net loss

No net movers

Not available

United States Census Bureau

Source: 2006-2010 5-year American Community Survey Estimates

Click to View County Statistics

Understanding the patterns of migration flows is important to government, business, and planners. The Census Flows Mapper shows the net gains and losses in population for the entire United States. The simple interface makes it easy for users to view, save, and print county-based maps.

http://flowsmapper.geo.census.gov/flowsmapper/flowsmapper.html

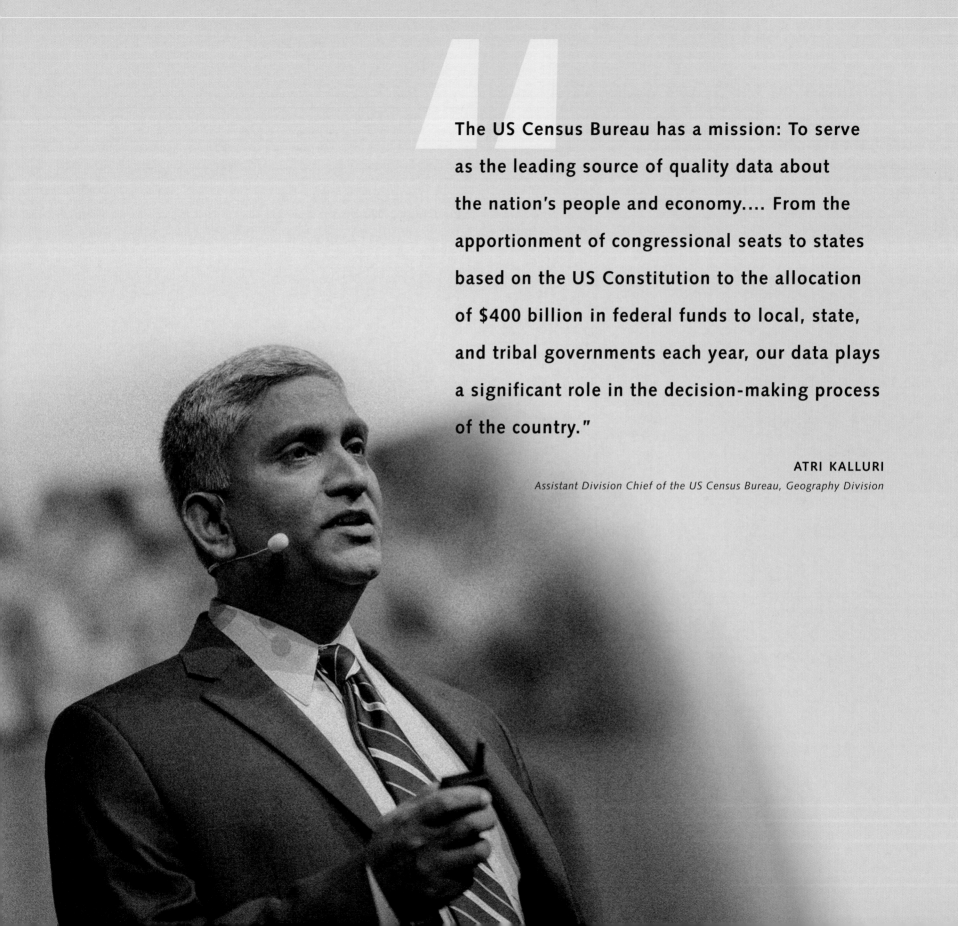

"The US Census Bureau has a mission: To serve as the leading source of quality data about the nation's people and economy.... From the apportionment of congressional seats to states based on the US Constitution to the allocation of $400 billion in federal funds to local, state, and tribal governments each year, our data plays a significant role in the decision-making process of the country."

ATRI KALLURI
Assistant Division Chief of the US Census Bureau, Geography Division

Census Tract Thematic Map Viewer

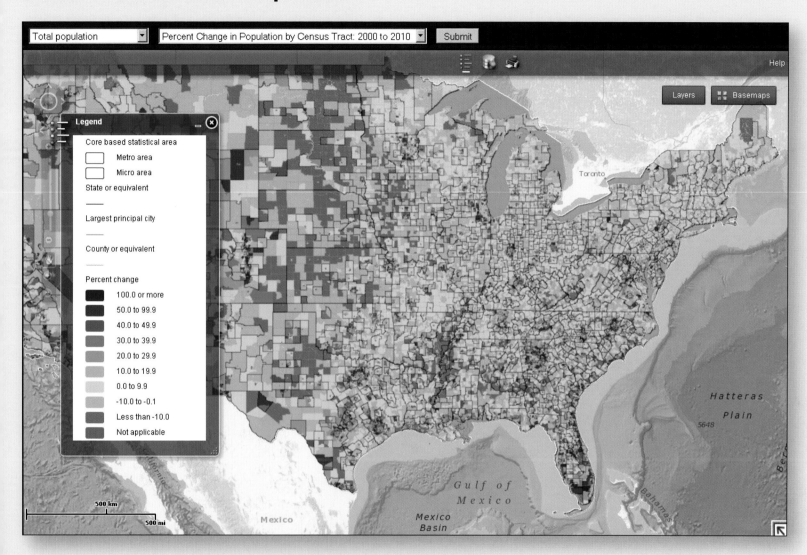

Knowing where people live and how populations have changed is necessary for planning and for many other decisions that must be made. Showing the percent change in population by census tract, this online application helps cut the cost of printing, improves the user experience, and enhances accessibility to the information.

 http://esriurl.com/7201

Ohio Congressional District 8 (Speaker John A. Boehner)

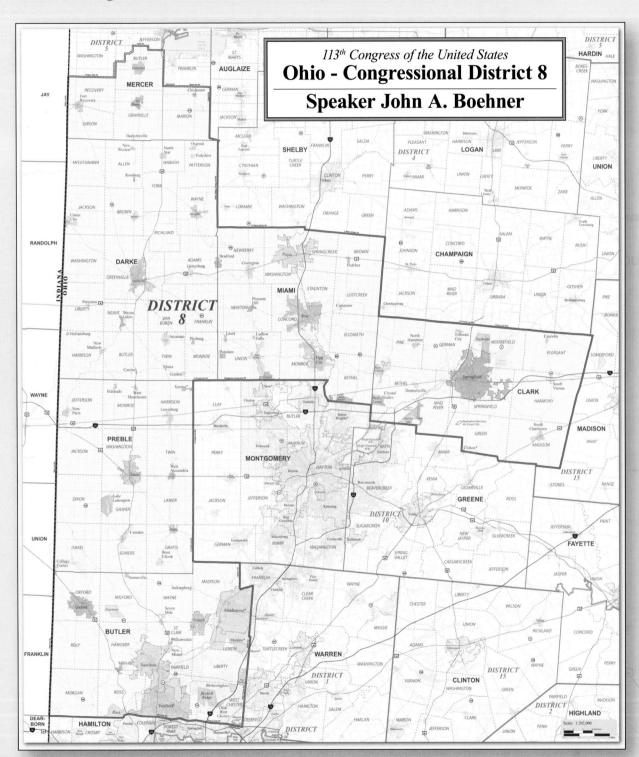

113th Congress of the United States
Ohio - Congressional District 8
Speaker John A. Boehner

Census data and mapping are essential to the redistricting process. The large-scale maps of the 113th Congress show boundaries as collected by the Redistricting Data Program. Such maps are an important part of helping to make public information accessible and easy to understand.

http://www.census.gov/rdo/

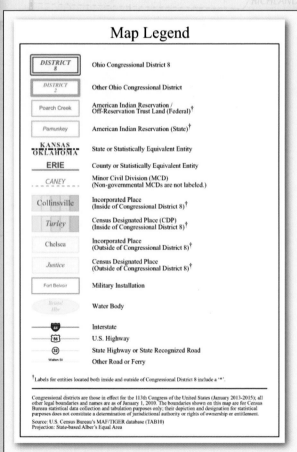

Map Legend

DISTRICT 8	Ohio Congressional District 8
DISTRICT 2	Other Ohio Congressional District
Poarch Creek	American Indian Reservation / Off-Reservation Trust Land (Federal)†
Pamunkey	American Indian Reservation (State)†
KANSAS OKLAHOMA	State or Statistically Equivalent Entity
ERIE	County or Statistically Equivalent Entity
CANEY	Minor Civil Division (MCD) (Non-governmental MCDs are not labeled.)
Collinsville	Incorporated Place (Inside of Congressional District 8)†
Turley	Census Designated Place (CDP) (Inside of Congressional District 8)†
Chelsea	Incorporated Place (Outside of Congressional District 8)†
Justice	Census Designated Place (Outside of Congressional District 8)†
Fort Belvoir	Military Installation
Bristol Blue	Water Body
	Interstate
	U.S. Highway
	State Highway or State Recognized Road
Walton St	Other Road or Ferry

†Labels for entities located both inside and outside of Congressional District 8 include a '*'.

Congressional districts are those in effect for the 113th Congress of the United States (January 2013-2015); all other legal boundaries and names are as of January 1, 2010. The boundaries shown on this map are for Census Bureau statistical data collection and tabulation purposes only; their depiction and designation for statistical purposes does not constitute a determination of jurisdictional authority or rights of ownership or entitlement.
Source: U.S. Census Bureau's MAF/TIGER database (TAB10)
Projection: State-based Alber's Equal Area

Congressional District 8 Profile

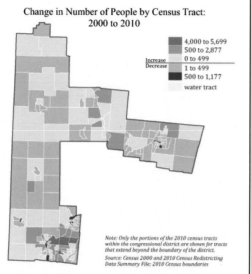

Change in Number of People by Census Tract:
2000 to 2010

Increase / Decrease
- 4,000 to 5,699
- 500 to 2,877
- 0 to 499
- 1 to 499
- 500 to 1,177
- water tract

Population
Total Population:	721,032
Age - Under 5:	*47,565*
Age - 5 to 17:	*130,855*
Age - 18 to 24:	*70,466*
Age - 25 to 39:	*128,593*
Age - 40 to 54:	*155,395*
Age - 55 to 64:	*89,768*
Age - 65+:	*98,390*
Percent Male:	49.0%
Percent Female:	51.0%
Persons per Square Mile:	294.2

Housing
Total Housing Units:	300,750
Percent Occupied:	91.5%
Percent Owner-Occupied:	*70.9%*
Percent Renter-Occupied:	*29.1%*
Percent Vacant:	8.5%

Land
Total Land Area (square miles):	2,450.49
Percent Land Area - Urban:	11.7%
Percent Land Area - Rural:	88.3%

Geographic Entity Tallies
American Indian Reservations/ Off-Reservation Trust Lands:	0 (0)
Counties:	6 (1)
Minor Civil Divisions (MCDs):	72 (1)
Incorporated Places:	71 (8)
Census Designated Places (CDPs):	13 (0)
ZIP Code Tabulation Areas:	88 (32)
Census Tracts:	173 (4)

***Total** (in bold) includes partial entities (in parentheses).*

Source: 2010 Census

Note: Only the portions of the 2010 census tracts within the congressional district are shown for tracts that extend beyond the boundary of the district.
Source: Census 2000 and 2010 Census Redistricting Data Summary File; 2010 Census boundaries

United States Census Bureau

For general information, contact the Congressional Affairs Office at (301) 763-6100. For more information regarding congressional district plans as a result of the 2010 Census, redistricting, and voting rights data, contact the Census Redistricting Data Office at (301) 763-4039 or www.census.gov/rdo. For information regarding other U.S. Census Bureau products, visit www.census.gov.

Location of Ohio's 8th Congressional District - 16 Districts Total

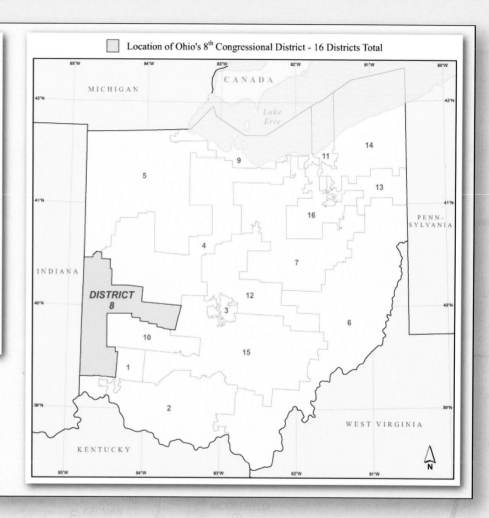

Congressional Districts of the 113th Congress of the United States (January 2013–2015)

Public access and knowledge play an important role in a democracy. This national map of the 113th Congress shows the boundaries of congressional districts based on the 2010 Census, and helps all citizens take an active role in government.

http://www.census.gov/rdo/

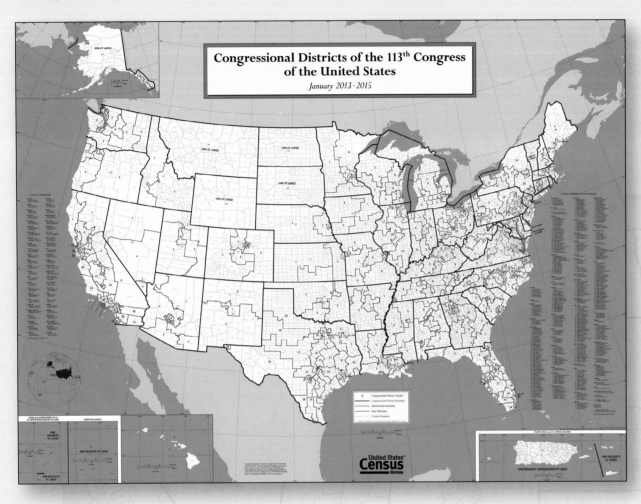

Small Area Income and Poverty Estimates

State	ID	Name	Estimate	Margin of Error (MOE)
AL	01001	Autauga County	14.9	2.4
AL	01003	Baldwin County	13.4	2.2
AL	01005	Barbour County	29.5	4.9
AL	01007	Bibb County	22.2	4.7
AL	01009	Blount County	14.9	3.0

Source: U.S. Census Bureau, Small Area Income and Poverty Estimates

The US Census Bureau is responsible for providing population data that is used in many areas of government. This interactive map allows states, counties, and school districts to explore annual income and poverty data quickly and easily. GIS enables formerly expensive and static printed maps to be replaced with interactive digital maps that are more usable and cost-efficient.

http://www.census.gov/population/metro/about/

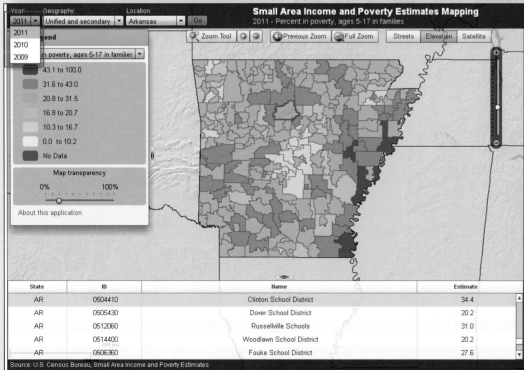

State	ID	Name	Estimate
AR	0504410	Clinton School District	34.4
AR	0505430	Dover School District	20.2
AR	0512060	Russellville Schools	31.0
AR	0514400	Woodlawn School District	20.2
AR	0506360	Fouke School District	27.6

Source: U.S. Census Bureau, Small Area Income and Poverty Estimates

US DEPARTMENT OF DEFENSE

The Battle of Wilson's Creek

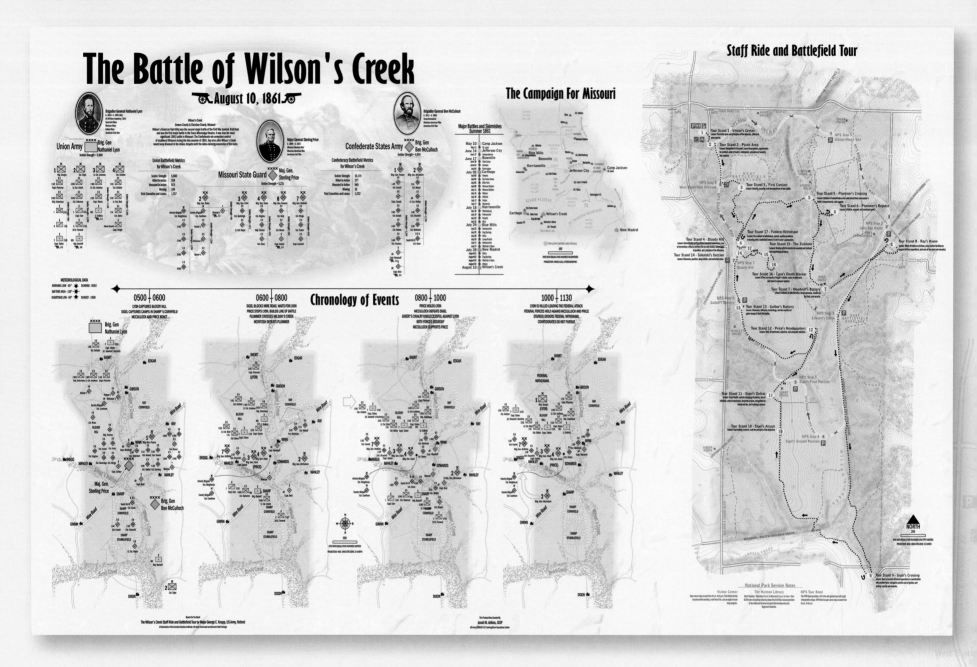

PROJECT BENEFITS

This training solution is valuable because the Army saves time and resources by giving officers the opportunity to take the virtual tour without visiting the actual site.

Valuable lessons can be gathered from past events such as those that occurred at Wilson's Creek, an important battle during the Civil War. The Combined Arms Center conducts staff rides and tours of the site so that soldiers can learn lessons that are still valuable to today's military leaders. The challenge was to enhance and update the training with a virtual tour. The Army accomplished this goal by creating digital maps that combine GIS and video game technology.

http://www.civilwarhome.com/wilsoncreekintro.htm

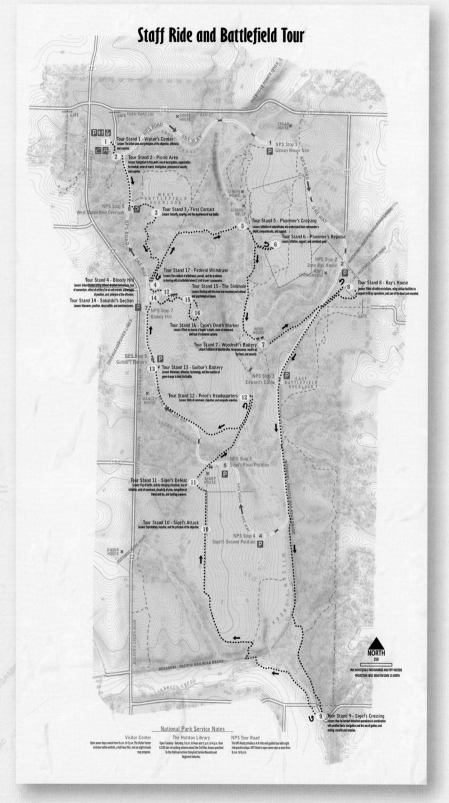

Arctic Sea Ice 2007–2012

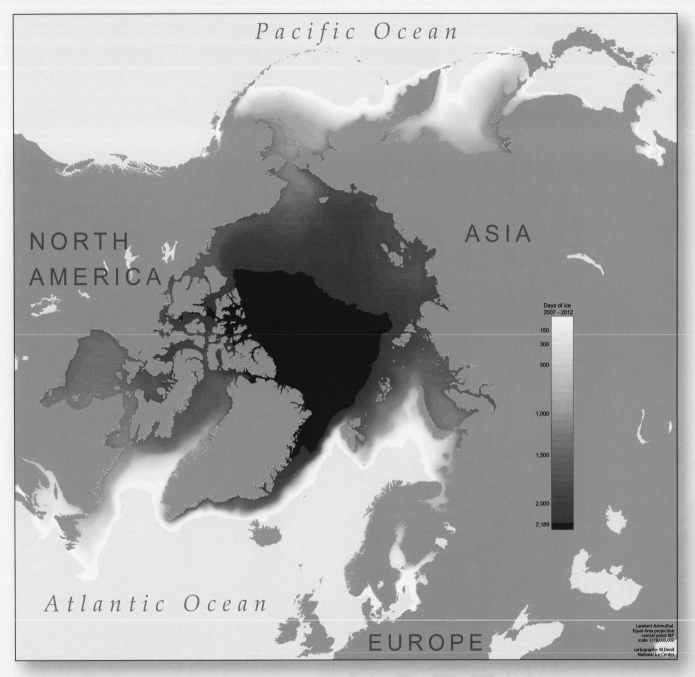

Mariners, scientists, and researchers need current and forecasted ice condition information for their work, and the US National Ice Center provides that information for both poles. This map compiles six years of US National Ice Center daily ice data. Maps such as these enable users to look at the history and state of Arctic sea ice, to help make predictions, and to assist in advancing the understanding of sea ice.

www.natice.noaa.gov

Days of ice
2007 - 2012

150
300
500
1,000
1,500
2,000
2,189

Lambert Azimuthal
Equal Area projection
central point: 90°
scale: 1/19,000,000

cartography: M.Denil
National Ice Center

2007

2008

2009

2010

2011

2012

Lambert Azimuthal
Equal Area projection
central point: 90°
scale: 1/90,700,000

ASIA

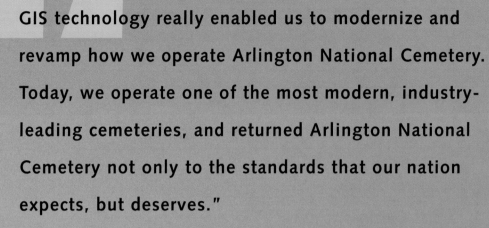

"GIS technology really enabled us to modernize and revamp how we operate Arlington National Cemetery. Today, we operate one of the most modern, industry-leading cemeteries, and returned Arlington National Cemetery not only to the standards that our nation expects, but deserves."

MAJOR NICHOLAS MILLER
Chief Information Officer of Arlington National Cemetery

Post-Superstorm Sandy Elevation Differences and Beach Volume Change

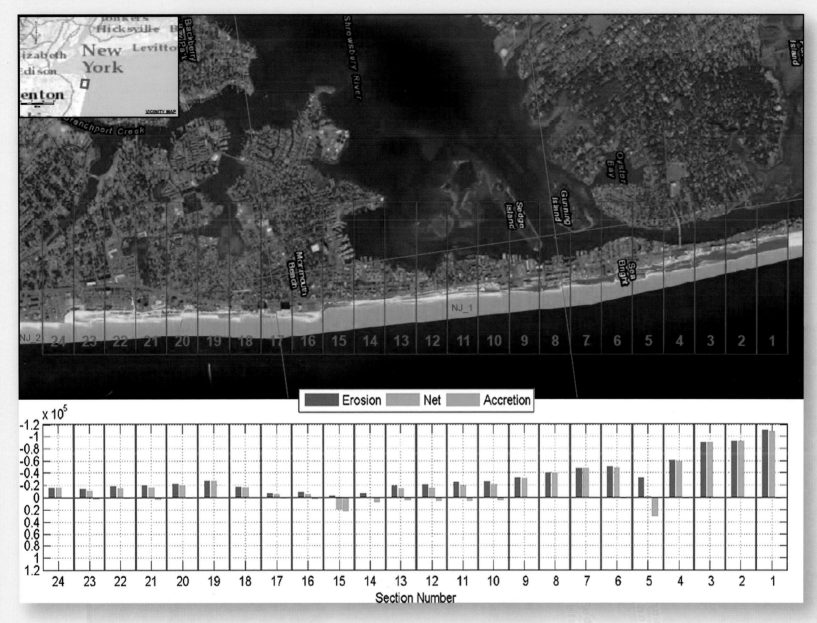

Aerial photography and map layers were combined to communicate the impacts of Hurricane Sandy on beaches in a concise way. The resulting map showed where erosion (brown) and buildup (blue) of sand took place. The bottom panel showed the amount of erosion and buildup that occurred in 1,000-foot, alongshore sections of the beach. The creation of this map clearly showed the alongshore variability in hurricane impacts on the beaches in the Northeast, and helped direct post-disaster response activities to mitigate the damages.

http://www.usace.army.mil/Missions/EmergencyOperations/ HurricaneSeason/Sandy.aspx

Drought Status in the Fort Worth District

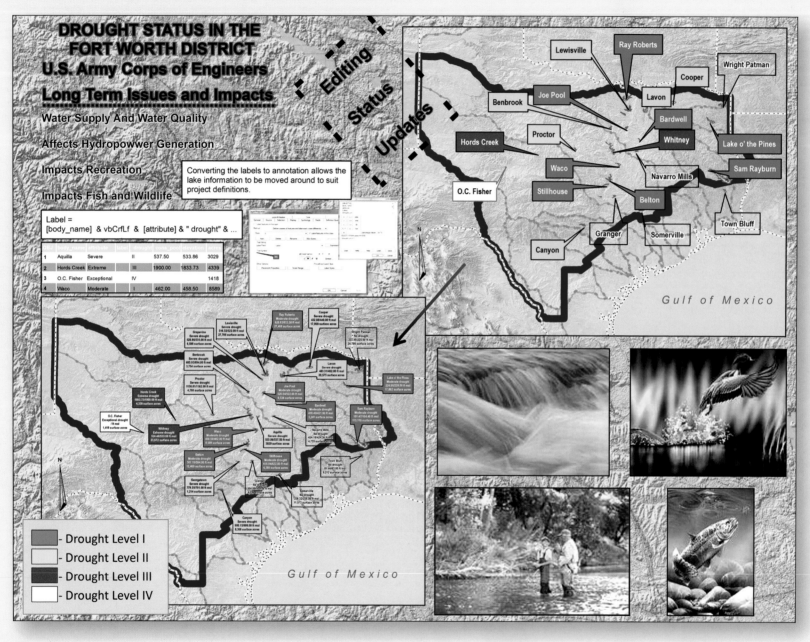

The US Army Corps of Engineers was faced with updating huge amounts of data to understand the impacts of a drought in Fort Worth, Texas. Officials required a way to plan for changes caused by the lack of rain. A GIS approach allowed analysts to update their information quickly and efficiently, saving them valuable time and resources. Government leaders, stakeholders, and the public had the resources necessary to understand the potential ramifications of the drought.

http://www.swf.usace.army.mil/

Extent of Flooding 1927 versus 2011

The US Army Corps of Engineers wanted to compare flooding on the lower Mississippi River in 2011 to the Great Mississippi flood of 1927. The goal was to understand the effectiveness of the levee system built after the earlier flood. The map shows that the 2011 flooding was lessened because of the levee system, potentially reducing property damage and loss of life.

 http://www.nwk.usace.army.mil/

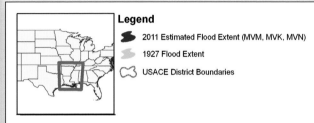

Legend

- 2011 Estimated Flood Extent (MVM, MVK, MVN)
- 1927 Flood Extent
- USACE District Boundaries

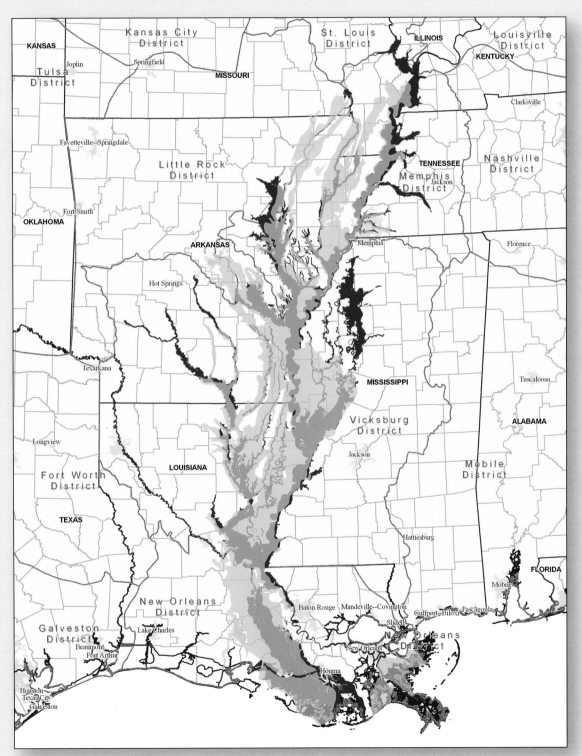

53

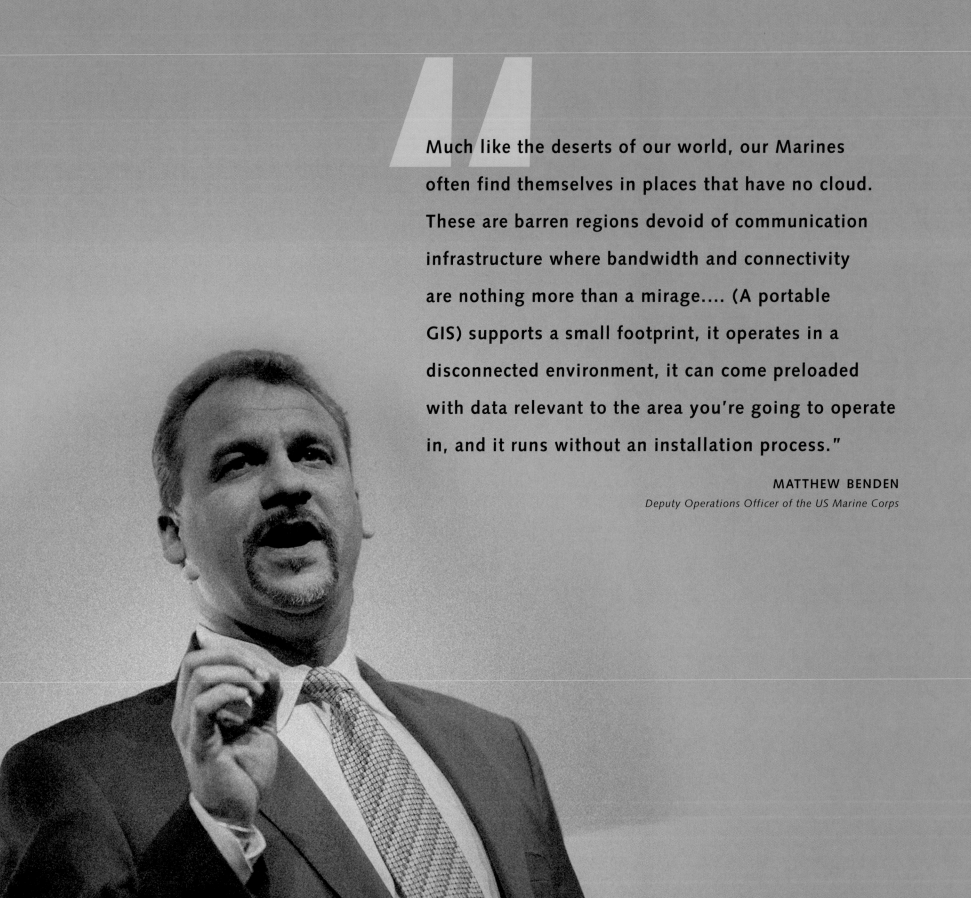

"Much like the deserts of our world, our Marines often find themselves in places that have no cloud. These are barren regions devoid of communication infrastructure where bandwidth and connectivity are nothing more than a mirage.... (A portable GIS) supports a small footprint, it operates in a disconnected environment, it can come preloaded with data relevant to the area you're going to operate in, and it runs without an installation process."

MATTHEW BENDEN
Deputy Operations Officer of the US Marine Corps

North Carolina National Guard Viewer

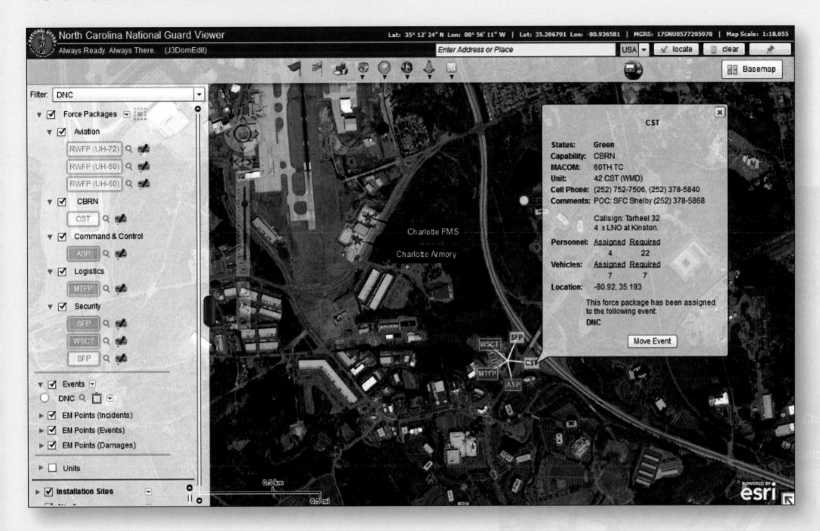

The North Carolina National Guard needed a way to show the status of a combat support team assigned to Camp Greene for the 2012 Democratic National Convention. They created an online map with GIS and aerial photography to achieve better efficiency and planning amongst the team. The venture also saved time and improved information sharing with real-time information going instantly to partnering agencies.

http://www.nc.ngb.army.mil/Pages/default.aspx

Navy Site Availability to Renewable Energy Resources

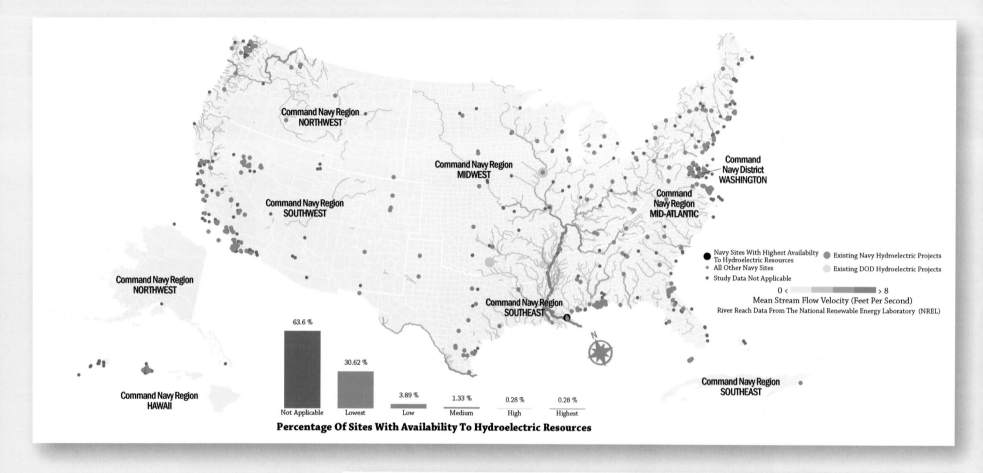

Navy Secretary Ray Mabus has set the Navy's goal of producing half of its energy from alternative sources by 2020. To help meet that goal, a series of maps was created to identify the Navy's renewable energy sites in the continental United States. These maps are a huge timesaver for policymakers who would otherwise have to search through multiple sources for the same information.

 http://www.navy.mil/local/cni/

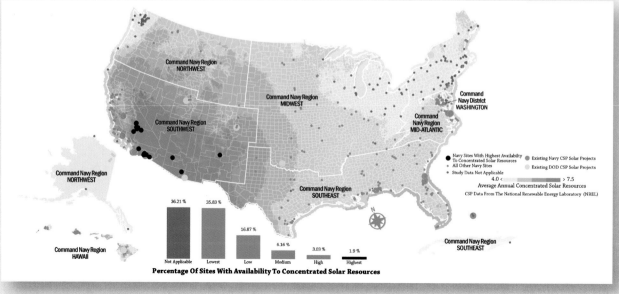

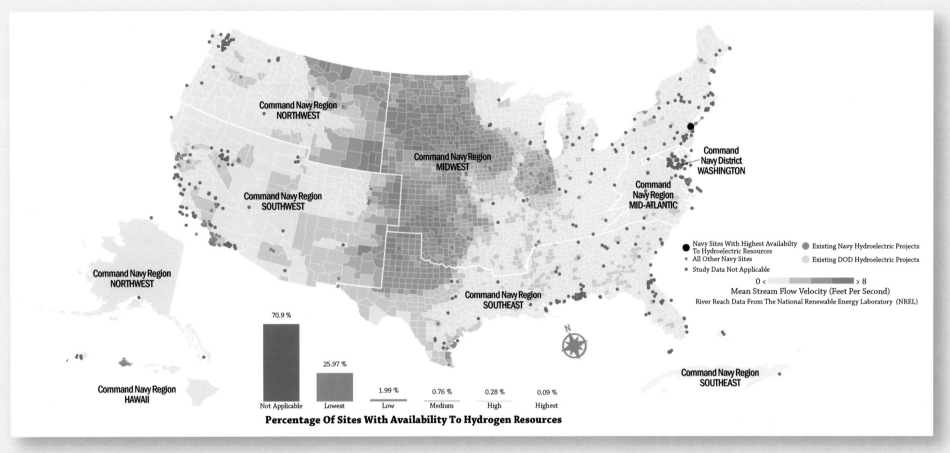

Percentage Of Sites With Availability To Hydrogen Resources

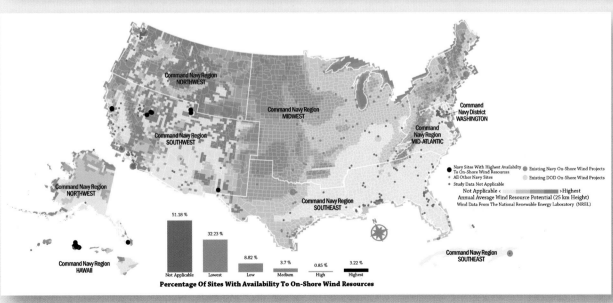

Percentage Of Sites With Availability To On-Shore Wind Resources

US DEPARTMENT OF HEALTH
AND HUMAN SERVICES

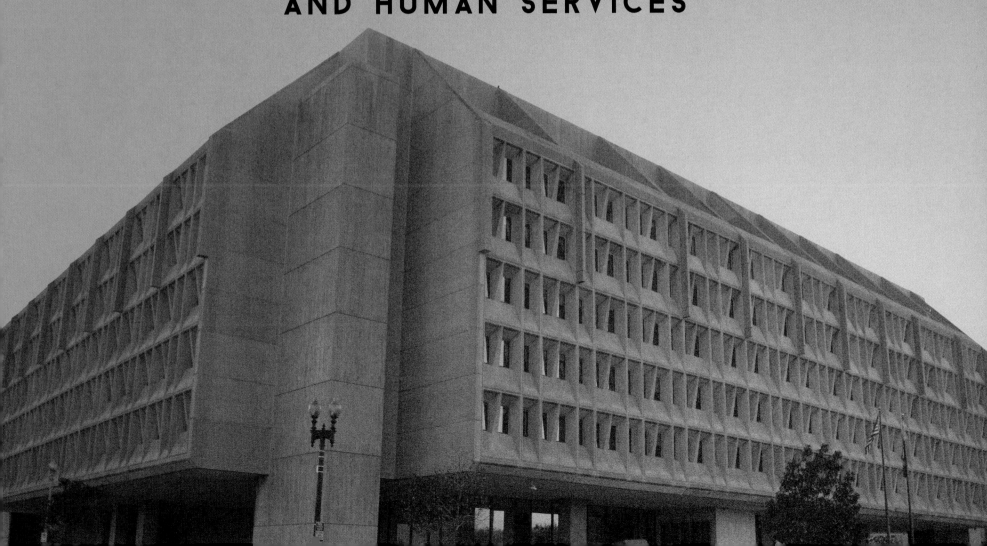

Validating the Need for Additional Federally Funded Health Centers

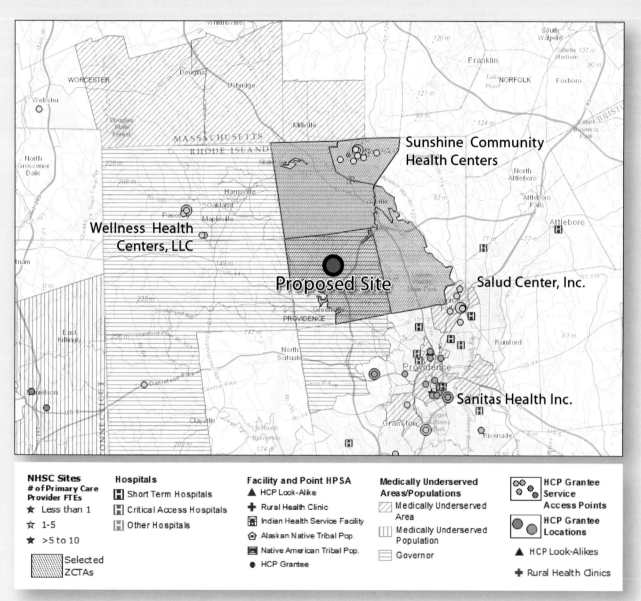

The Health Resources and Services Administration (HRSA) is charged with improving access to healthcare services for people who are uninsured, isolated, or medically vulnerable. To accomplish that goal, the HRSA offers grants to providers who create new primary healthcare services in medically underserved communities. The UDS Mapper is a valuable tool for validating the need for additional federally funded health centers and ensuring that funding is directed where needed, without service area overlap.

http://www.hrsa.gov/about/index.html

NHSC Sites # of Primary Care Provider FTEs	Hospitals	Facility and Point HPSA	Medically Underserved Areas/Populations	HCP Grantee Service Access Points
★ Less than 1	H Short Term Hospitals	▲ HCP Look-Alike	▨ Medically Underserved Area	
☆ 1-5	H Critical Access Hospitals	✚ Rural Health Clinic	⫴ Medically Underserved Population	HCP Grantee Locations
★ >5 to 10	H Other Hospitals	▦ Indian Health Service Facility		
		⌂ Alaskan Native Tribal Pop.	▤ Governor	▲ HCP Look-Alikes
▨ Selected ZCTAs		▣ Native American Tribal Pop.		✚ Rural Health Clinics
		● HCP Grantee		

Health Professional Shortage Areas—Primary Care

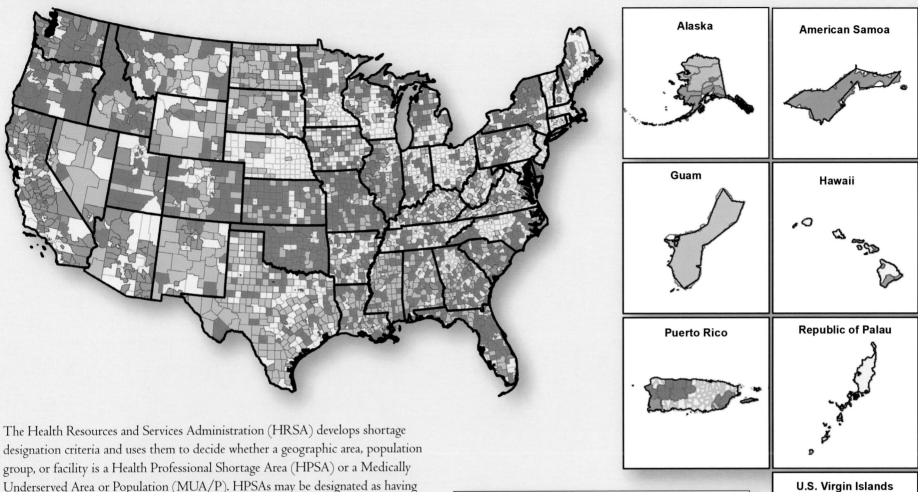

The Health Resources and Services Administration (HRSA) develops shortage designation criteria and uses them to decide whether a geographic area, population group, or facility is a Health Professional Shortage Area (HPSA) or a Medically Underserved Area or Population (MUA/P). HPSAs may be designated as having a shortage of primary medical care, dental, or mental health providers. HRSA used GIS to create this map of designated Primary Care HPSAs. Investment in GIS technologies has improved operating efficiency, data quality, customer impact/ burden, and service availability since 2003.

http://www.hrsa.gov/about/index.html

Legend

Health Professional Shortage Areas - Primary Care

- Geographical Area
- Population Group
- Single County
- Not Primary Care HPSA

Global Spread of Disease Caused by International Travel

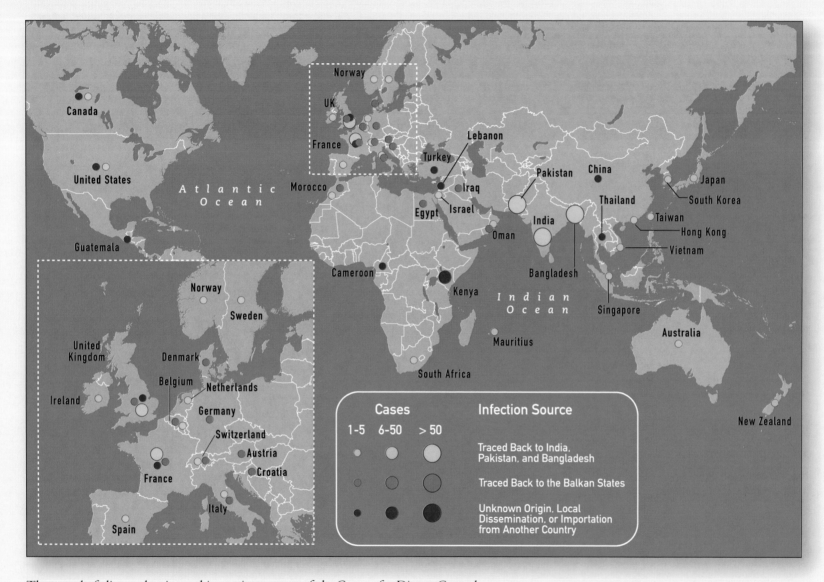

The spread of diseases by air travel is a major concern of the Centers for Disease Control and Prevention. An important goal of the agency is to create digital maps that can help trace the origin of disease, identify the number of infections, and track its spread. This map identifies the number of cases of an infection caused by international travel.

http://wwwnc.cdc.gov/eid/article/17/10/11-0655_article.htm

Interactive Atlas of Heart Disease and Stroke

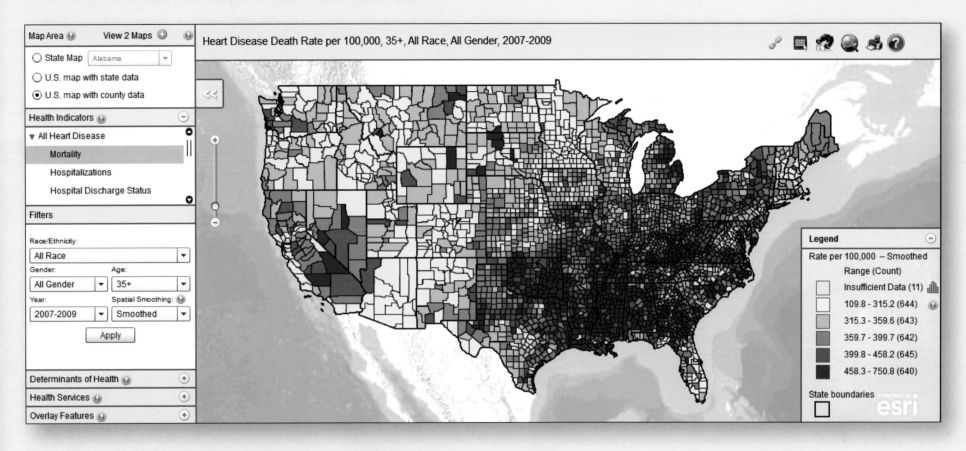

The Centers for Disease Control and Prevention needed to find a way to share maps of heart disease and stroke with public health professionals working at the state and county levels. To answer this challenge, the Interactive Atlas of Heart Disease and Stroke was created. As a result, state and local health departments, hospitals, community organizations, and academic centers can generate expansive sets of maps that enable them to tailor heart disease and stroke prevention programs and policies to the needs of specific counties.

 http://apps.nccd.cdc.gov/dhdspatlas/

63

Determining Safe Shellfish Growing Areas

Blaine, Washington

Actual Outfall
Shellfish Cages

**Track Data
Concentration ppb**
<= 0.01
0.01 to 0.05
0.05 to 0.10
0.10 to 0.50
0.50 to 1.0
1.0 to 5.0
5.0 to 10
10 to 50
> 50

The timely compilation and evaluation of data is important for food safety and nutrition. It had previously taken months to do this for shellfish growing areas. The Food and Drug Administration's Center for Food Safety and Applied Nutrition developed the Real-Time Application for Tracking and Mapping (RAFT-MAP). Built on custom Esri technology, this system enables shellfish growing area evaluations to be performed in weeks instead of months, saving the government time and money.

http://www.fda.gov/Food/default.htm

Animated Historical Cancer Atlas

The National Cancer Institute, part of the National Institutes of Health and the Department of Health and Human Services, needed historical patterns of cancer mortality from 1971 to 2010 for supporting cancer research. Age-adjusted deaths per 100,000 are computed, modeled, and presented in five-year intervals. Blue represents the lowest death rates, and red represents the highest. Using GIS to analyze and present information helps researchers to model and examine patterns to aid them in their work.

 http://ratecalc.cancer.gov/

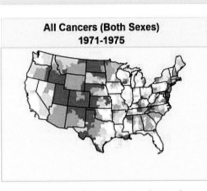

All Cancers (Both Sexes) 1971-1975

1971–1975 ⬇ Text View | PDF | JPG

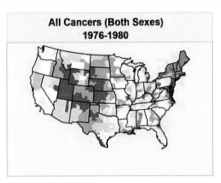

All Cancers (Both Sexes) 1976-1980

1976–1980 ⬇ Text View | PDF | JPG

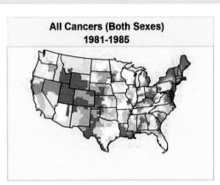

All Cancers (Both Sexes) 1981-1985

1981–1985 ⬇ Text View | PDF | JPG

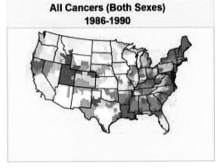

All Cancers (Both Sexes) 1986-1990

1986–1990 ⬇ Text View | PDF | JPG

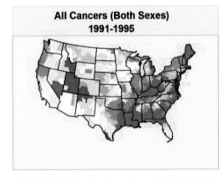

All Cancers (Both Sexes) 1991-1995

1991–1995 ⬇ Text View | PDF | JPG

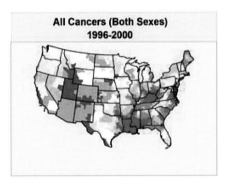

All Cancers (Both Sexes) 1996-2000

1996–2000 ⬇ Text View | PDF | JPG

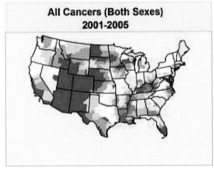

All Cancers (Both Sexes) 2001-2005

2001–2005 ⬇ Text View | PDF | JPG

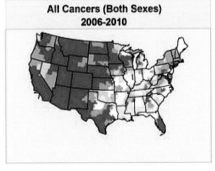

All Cancers (Both Sexes) 2006-2010

2006–2010 ⬇ Text View | PDF | JPG

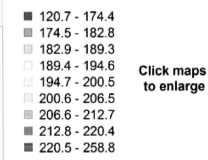

■ 120.7 - 174.4
▨ 174.5 - 182.8
▦ 182.9 - 189.3
□ 189.4 - 194.6
▢ 194.7 - 200.5
▨ 200.6 - 206.5
▨ 206.6 - 212.7
■ 212.8 - 220.4
■ 220.5 - 258.8

Click maps to enlarge

(Deaths per 100,000)

US DEPARTMENT OF HOMELAND SECURITY

Data is all around us and the best way to make operational decisions is by displaying it on GIS. We are literally entering into a golden age of information sharing.... The technology exists that allows common operating platforms to be easily configured to meet a spectrum of different agency needs and missions."

ROBERT GRIFFIN
Director, First Responders Division, US Department of Homeland Security

Fifth District SAR Cases Density Plot

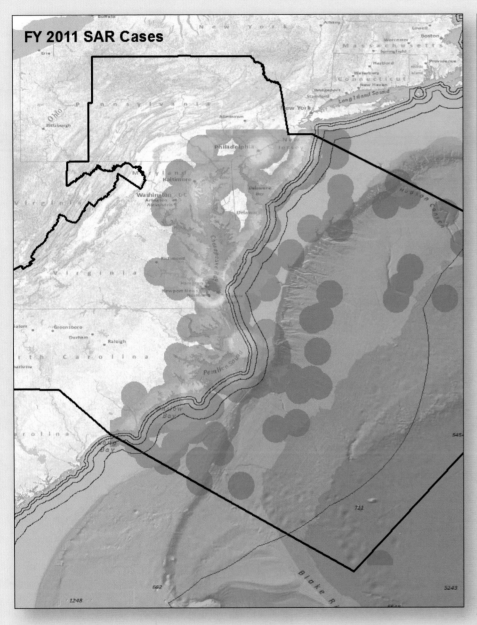

FY 2011 SAR Cases

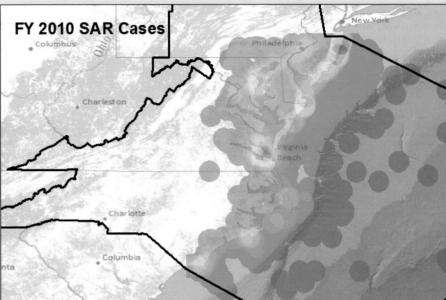

FY 2010 SAR Cases

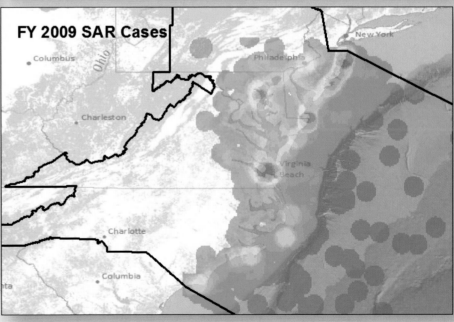

FY 2009 SAR Cases

Examining patterns in incident locations can help to illuminate emergency situations. This GIS map displays search and rescue cases of the Fifth Coast Guard District in the Mid-Atlantic region for three fiscal years. The information it presents can be used to examine possible trends or causes and thereby help in future search and rescue operations.

http://www.uscg.mil/d5/

Winter Storm Nemo Forecast Impacts

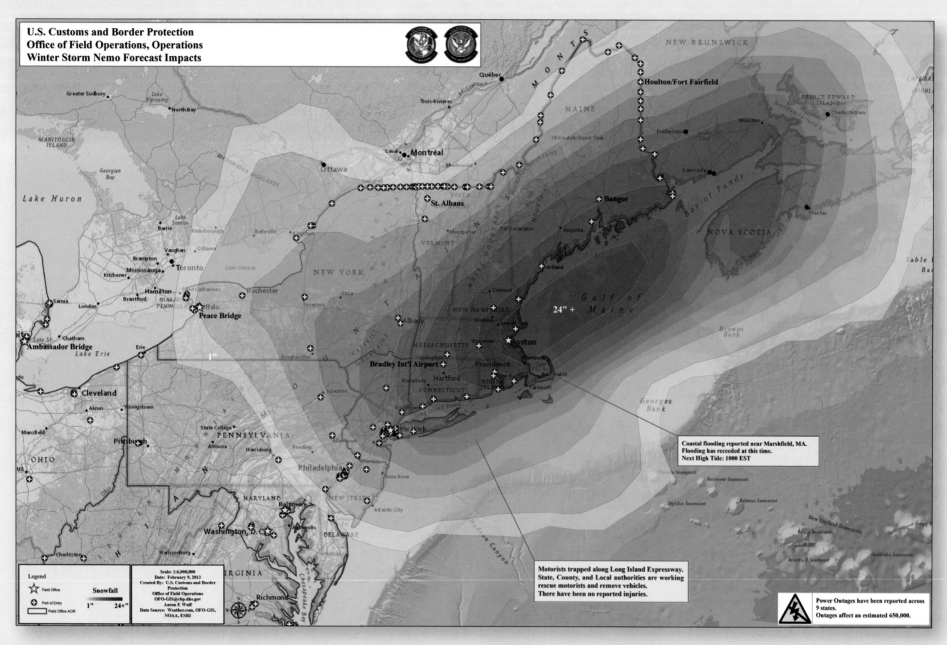

U.S. Customs and Border Protection
Office of Field Operations, Operations
Winter Storm Nemo Forecast Impacts

Coastal flooding reported near Marshfield, MA.
Flooding has receded at this time.
Next High Tide: 1000 EST

Motorists trapped along Long Island Expressway.
State, County, and Local authorities are working
rescue motorists and remove vehicles.
There have been no reported injuries.

Power Outages have been reported across
9 states.
Outages affect an estimated 650,000.

Legend

Field Office
Port of Entry
Field Office AOR

Snowfall
1"   24"+

Scale: 1:6,000,000
Date: February 9, 2013
Created By: U.S. Customs and Border
Protection
Office of Field Operations
OFO-GIS@cbp.dhs.gov
Aaron F. Wolf
Data Source: Weather.com, OFO-GIS,
NOAA, ESRI

Winter Storm Nemo brought the potential for record-breaking snowfall to New
England and affected US Customs and Border Protection (CBP) operations. This map
depicts the snowfall threat and identifies field operations to answer questions about
which locations would fare worse. Emergency preparedness is a crucial component of
border security and movement of goods.

http://www.cbp.gov/

US Southwest Border

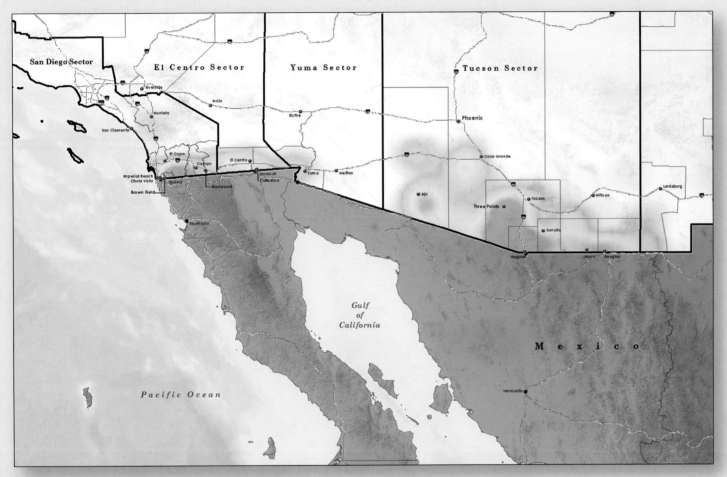

Ensuring border security and preventing illegal immigration are important functions of the US Border Patrol. Agents use GIS to analyze border activity and improve border management, which can give decision makers the information needed to better allocate resources.

http://www.cbp.gov/xp/cgov/border_security/border_patrol/

US DEPARTMENT OF HOUSING
AND URBAN DEVELOPMENT

Sandy Damage Estimates by Block Group

After a large-scale natural disaster, it is important to provide assistance to victims as quickly as possible. This map shows how housing inspection data was analyzed by the Federal Emergency Management Agency (FEMA) to assess property and infrastructure damage following Hurricane Sandy and to determine assistance eligibility for storm survivors. In cases like this, GIS enables FEMA and HUD to help people more efficiently.

http://www.huduser.org/maps/map_sandy_blockgroup.html

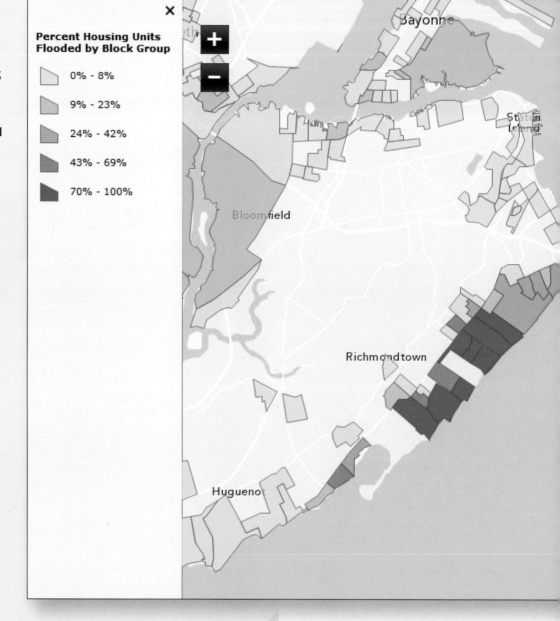

Percent Housing Units
Flooded by Block Group

0% - 8%

9% - 23%

24% - 42%

43% - 69%

70% - 100%

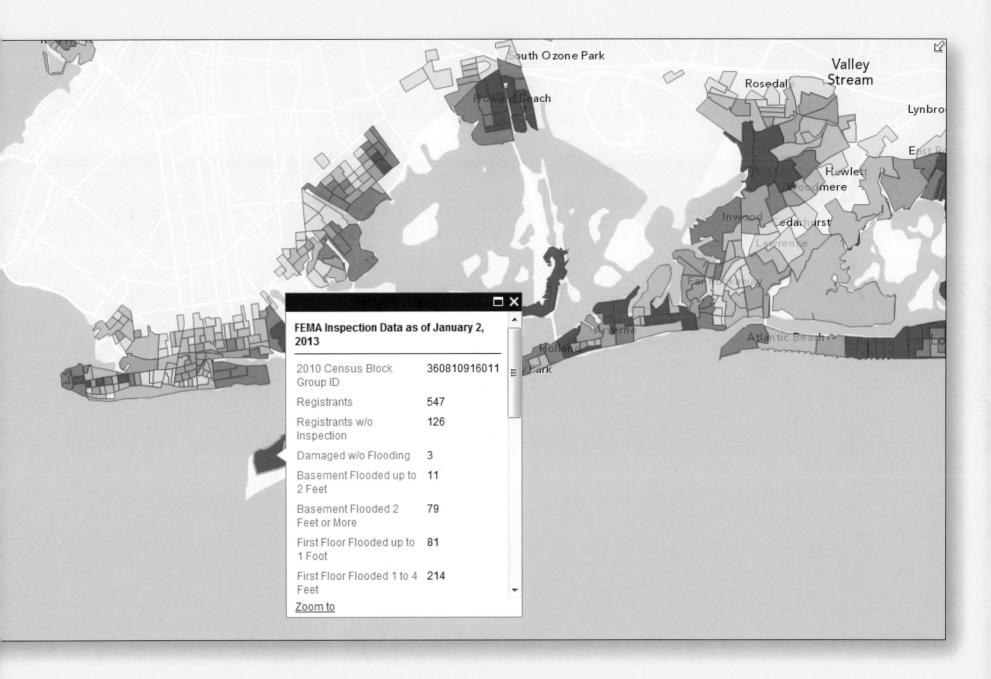

FEMA Inspection Data as of January 2, 2013

2010 Census Block Group ID	360810916011
Registrants	547
Registrants w/o Inspection	126
Damaged w/o Flooding	3
Basement Flooded up to 2 Feet	11
Basement Flooded 2 Feet or More	79
First Floor Flooded up to 1 Foot	81
First Floor Flooded 1 to 4 Feet	214

Zoom to

US DEPARTMENT OF THE INTERIOR

"The need to effectively leverage and share each other's work is becoming a bigger and bigger driver in each of our organizations every day. In a recent study, GAO (Government Accountability Office) emphasized the need for the federal geospatial community to work more closely together and to share our data and tools across our organizational boundaries."

JERRY JOHNSON
Geospatial Information Officer of the US Department of the Interior

Fort Ord National Monument Trail Map

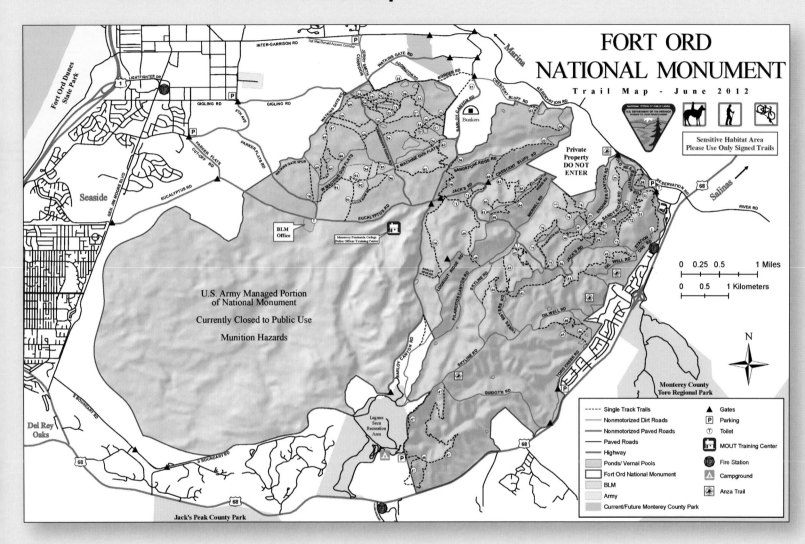

Restoration work is constantly under way at Fort Ord National Monument, which means the trails for public hiking and riding are regularly being rerouted. To make sure federal lands are accessible during the overhaul, the Bureau of Land Management created this interim map to show the location of available trails and access points to the monument. Though always subject to change, these trail maps allow the bureau to provide recreational activities to the public while ensuring that visitors feel confident to explore the area safely despite the constant renovations.

http://www.blm.gov/pgdata/content/ca/en/fo/hollister/fort_ord/index.html

North Half Oregon Central Coast Recreation Map

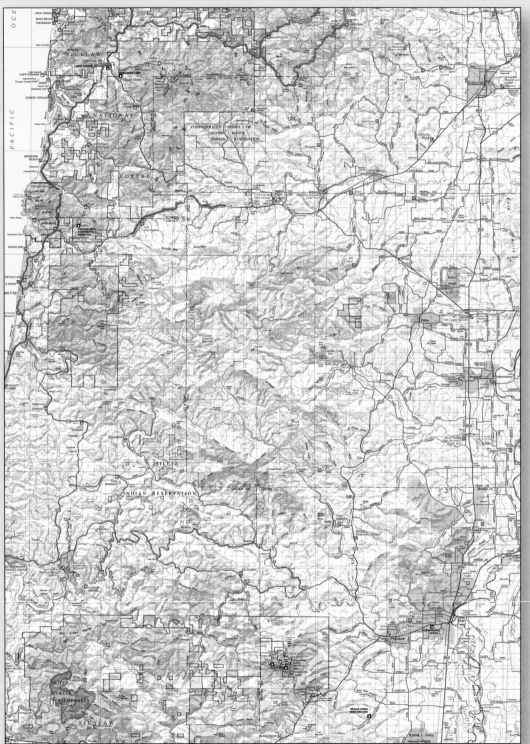

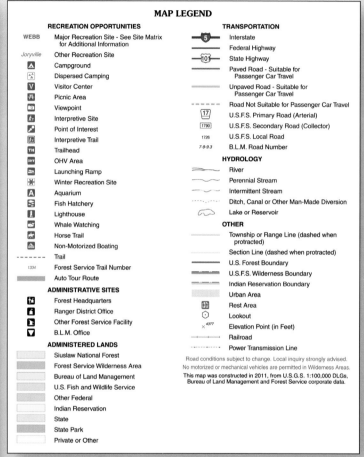

MAP LEGEND

RECREATION OPPORTUNITIES

WEBB	Major Recreation Site - See Site Matrix for Additional Information
Joryville	Other Recreation Site
	Campground
	Dispersed Camping
	Visitor Center
	Picnic Area
	Viewpoint
	Interpretive Site
	Point of Interest
	Interpretive Trail
TH	Trailhead
OHV	OHV Area
	Launching Ramp
	Winter Recreation Site
	Aquarium
	Fish Hatchery
	Lighthouse
	Whale Watching
	Horse Trail
	Non-Motorized Boating
	Trail
1334	Forest Service Trail Number
	Auto Tour Route

ADMINISTRATIVE SITES

	Forest Headquarters
	Ranger District Office
	Other Forest Service Facility
	B.L.M. Office

ADMINISTERED LANDS

- Siuslaw National Forest
- Forest Service Wilderness Area
- Bureau of Land Management
- U.S. Fish and Wildlife Service
- Other Federal
- Indian Reservation
- State
- State Park
- Private or Other

TRANSPORTATION

5	Interstate
	Federal Highway
101	State Highway
	Paved Road - Suitable for Passenger Car Travel
	Unpaved Road - Suitable for Passenger Car Travel
	Road Not Suitable for Passenger Car Travel
17	U.S.F.S. Primary Road (Arterial)
1790	U.S.F.S. Secondary Road (Collector)
1726	U.S.F.S. Local Road
7-9-9.3	B.L.M. Road Number

HYDROLOGY

	River
	Perennial Stream
	Intermittent Stream
	Ditch, Canal or Other Man-Made Diversion
	Lake or Reservoir

OTHER

	Township or Range Line (dashed when protracted)
	Section Line (dashed when protracted)
	U.S. Forest Boundary
	U.S.F.S. Wilderness Boundary
	Indian Reservation Boundary
	Urban Area
	Rest Area
	Lookout
×4377	Elevation Point (in Feet)
	Railroad
	Power Transmission Line

Road conditions subject to change. Local inquiry strongly advised.
No motorized or mechanical vehicles are permitted in Wilderness Areas.
This map was constructed in 2011, from U.S.G.S. 1:100,000 DLGs, Bureau of Land Management and Forest Service corporate data.

Oregon has a wide range of recreational opportunities for outdoor enthusiasts. This map is part of a multiagency map series covering the Pacific Northwest. Maps in this series are for sale to the public and advertise the recreational features and opportunities throughout Oregon and Washington. This high quality map contains valuable information and covers a wide range of recreational interests.

http://www.blm.gov/or/onlineservices/maps/map_details.php?id=28

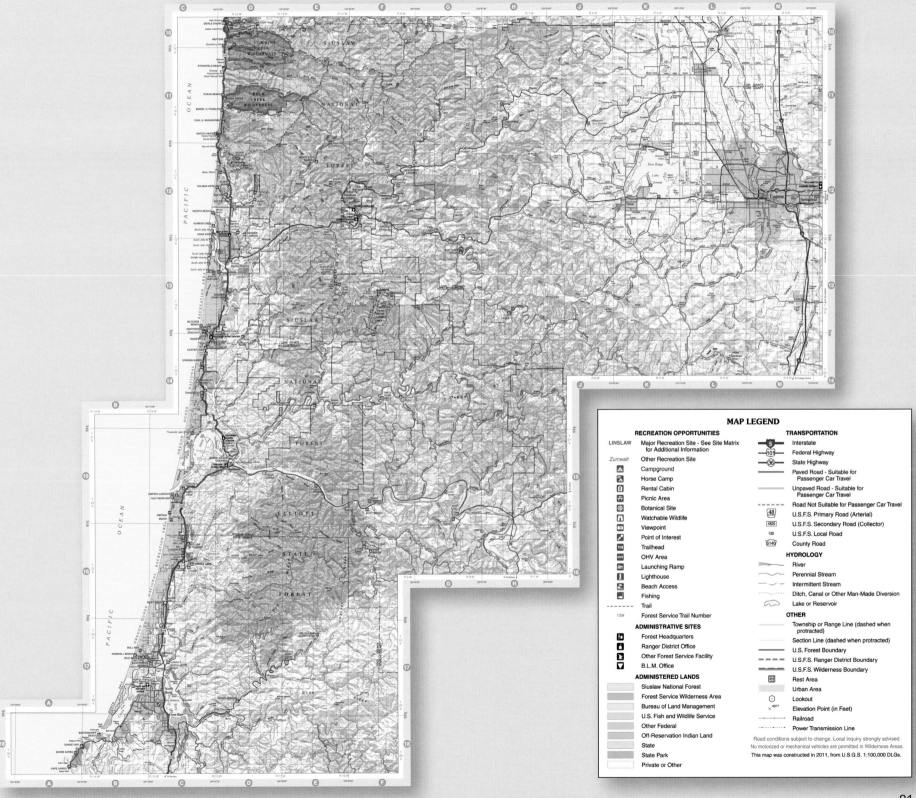

MAP LEGEND

RECREATION OPPORTUNITIES

LINSLAW — Major Recreation Site - See Site Matrix for Additional Information

Zumwalt — Other Recreation Site

- Campground
- Horse Camp
- Rental Cabin
- Picnic Area
- Botanical Site
- Watchable Wildlife
- Viewpoint
- Point of Interest
- Trailhead
- OHV Area
- Launching Ramp
- Lighthouse
- Beach Access
- Fishing
- Trail
- 1234 — Forest Service Trail Number

ADMINISTRATIVE SITES

- Forest Headquarters
- Ranger District Office
- Other Forest Service Facility
- B.L.M. Office

ADMINISTERED LANDS

- Siuslaw National Forest
- Forest Service Wilderness Area
- Bureau of Land Management
- U.S. Fish and Wildlife Service
- Other Federal
- Off-Reservation Indian Land
- State
- State Park
- Private or Other

TRANSPORTATION

- 5 — Interstate
- 101 — Federal Highway
- 36 — State Highway
- Paved Road - Suitable for Passenger Car Travel
- Unpaved Road - Suitable for Passenger Car Travel
- Road Not Suitable for Passenger Car Travel
- 48 — U.S.F.S. Primary Road (Arterial)
- 4820 — U.S.F.S. Secondary Road (Collector)
- 130 — U.S.F.S. Local Road
- 5140 — County Road

HYDROLOGY

- River
- Perennial Stream
- Intermittent Stream
- Ditch, Canal or Other Man-Made Diversion
- Lake or Reservoir

OTHER

- Township or Range Line (dashed when protracted)
- Section Line (dashed when protracted)
- U.S. Forest Boundary
- U.S.F.S. Ranger District Boundary
- U.S.F.S. Wilderness Boundary
- Rest Area
- Urban Area
- Lookout
- ×4377 — Elevation Point (in Feet)
- Railroad
- Power Transmission Line

Road conditions subject to change. Local inquiry strongly advised.
No motorized or mechanical vehicles are permitted in Wilderness Areas.
This map was constructed in 2011, from U.S.G.S. 1:100,000 DLGs,

San Juan Islands—Bureau of Land Management Jurisdiction

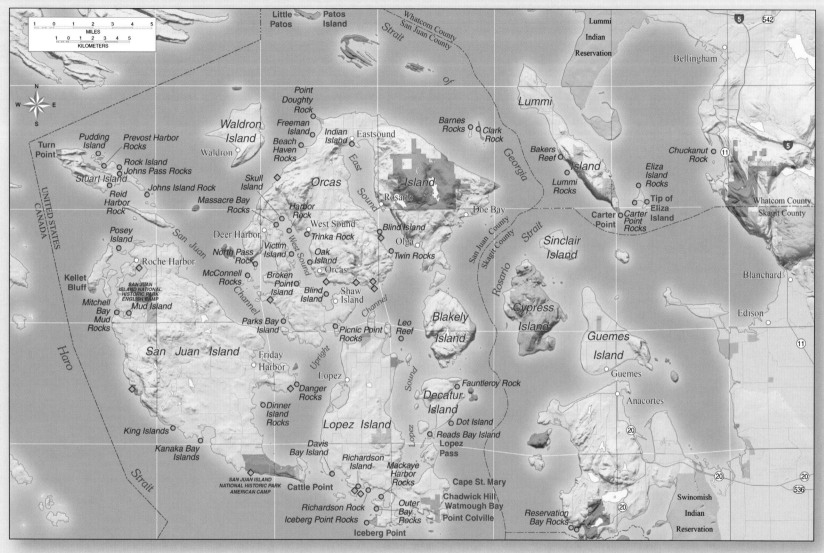

The Bureau of Land Management (BLM) is tasked with preserving and protecting cultural, ecological, and scientifically valuable areas for the benefit of future generations. This map uses GIS to show which BLM islands in Puget Sound, Washington, are being considered for congressional protection. This map gives legislators the information they need to support their decisions.

http://www.sanjuanislandsnca.org/blm-lands-in-the-san-juans

BLM Geoportal and Map Viewer

The Bureau of Land Management supports open government initiatives through the sharing and distribution of data. This GIS portal is designed with broad customer benefit in mind to provide sharing, previewing, browsing, searching, and discovery of information at low cost. The use of a portal allows the BLM to upload data once and distribute that data across numerous publishing platforms.

http://www.blm.gov/or/index.php

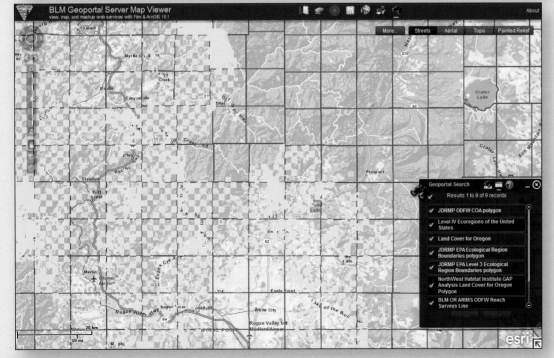

Reservations with Significant Timberland Resources

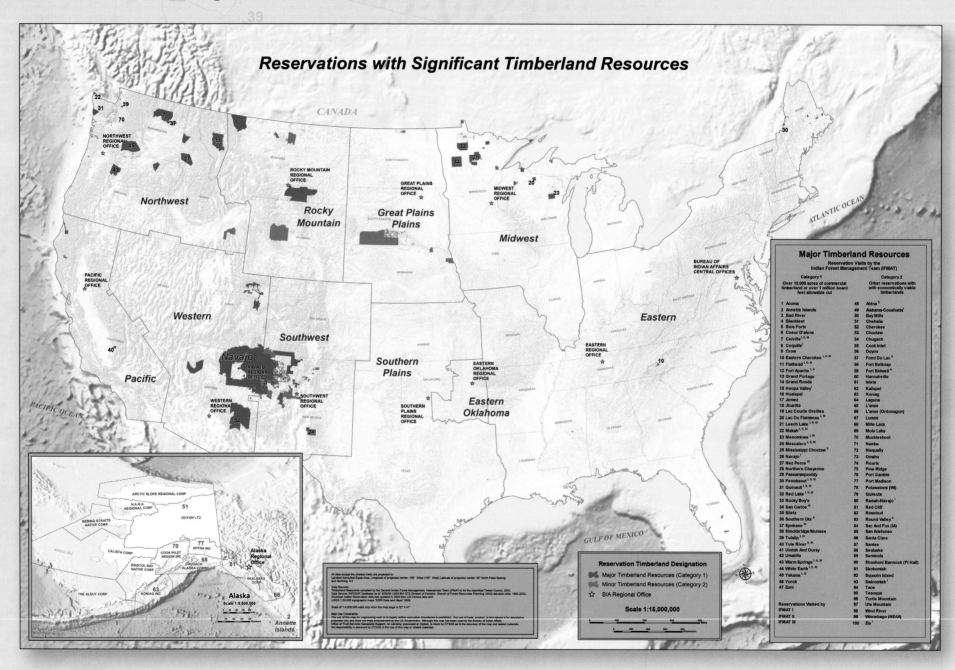

The Indian Forest Management Assessment Team (IFMAT) manages forest resources and convenes every ten years to provide an independent status assessment. IFMAT is mandated by the National Indian Forest Resources Management Act. These teams collected data and with GIS revealed that reservations have significant forest resources.

In addition to identifying those reservations, the map provides a wealth of information that is currently being used by Bureau of Indian Affairs in the forestry field.

http://www.bia.gov/WhoWeAre/BIA/OTS/DFWFM/NIFP/

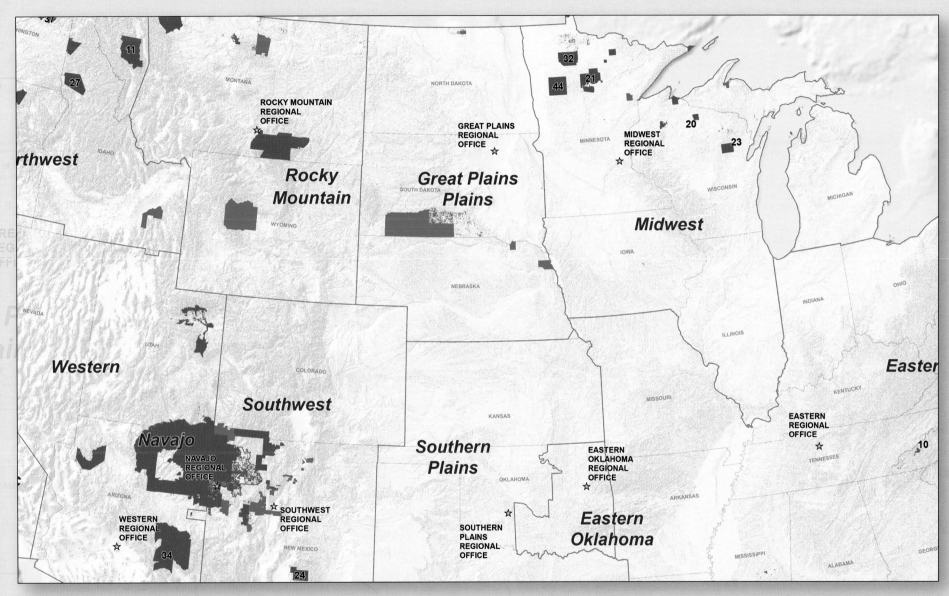

85

The Hazard of Melting Permafrost for the Alaskan Natives

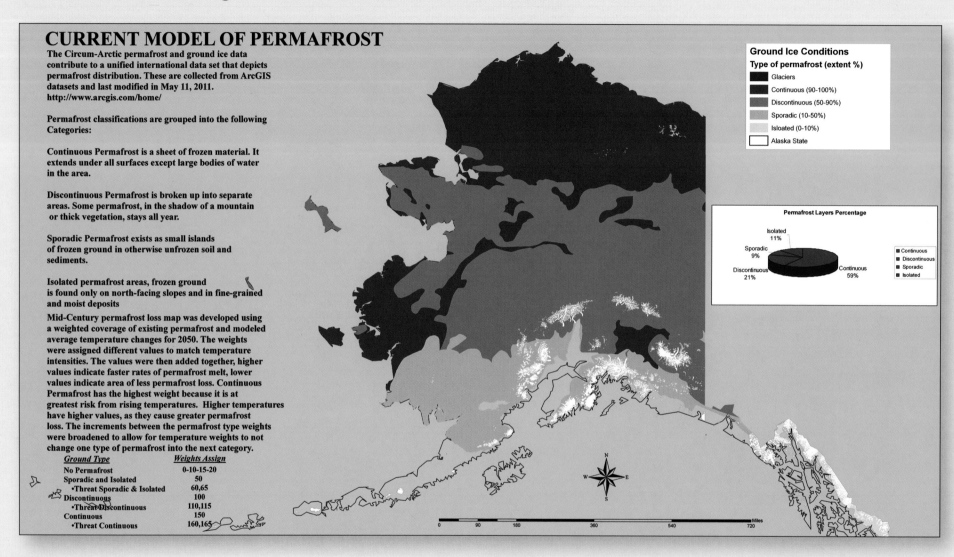

CURRENT MODEL OF PERMAFROST

The Circum-Arctic permafrost and ground ice data contribute to a unified international data set that depicts permafrost distribution. These are collected from ArcGIS datasets and last modified in May 11, 2011.
http://www.arcgis.com/home/

Permafrost classifications are grouped into the following Categories:

Continuous Permafrost is a sheet of frozen material. It extends under all surfaces except large bodies of water in the area.

Discontinuous Permafrost is broken up into separate areas. Some permafrost, in the shadow of a mountain or thick vegetation, stays all year.

Sporadic Permafrost exists as small islands of frozen ground in otherwise unfrozen soil and sediments.

Isolated permafrost areas, frozen ground is found only on north-facing slopes and in fine-grained and moist deposits

Mid-Century permafrost loss map was developed using a weighted coverage of existing permafrost and modeled average temperature changes for 2050. The weights were assigned different values to match temperature intensities. The values were then added together, higher values indicate faster rates of permafrost melt, lower values indicate area of less permafrost loss. Continuous Permafrost has the highest weight because it is at greatest risk from rising temperatures. Higher temperatures have higher values, as they cause greater permafrost loss. The increments between the permafrost type weights were broadened to allow for temperature weights to not change one type of permafrost into the next category.

Ground Type	Weights Assign
No Permafrost	0-10-15-20
Sporadic and Isolated	50
•Threat Sporadic & Isolated	60,65
Discontinuous	100
•Threat Discontinuous	110,115
Continuous	150
•Threat Continuous	160,165

Ground Ice Conditions

Type of permafrost (extent %)

- Glaciers
- Continuous (90-100%)
- Discontinuous (50-90%)
- Sporadic (10-50%)
- Isloated (0-10%)
- Alaska State

Permafrost Layers Percentage

- Isolated 11%
- Sporadic 9%
- Discontinuous 21%
- Continuous 59%

■ Continuous
■ Discontinuous
■ Sporadic
■ Isolated

Tribal communities in Alaska rely on permanently frozen ground conditions for survival. A GIS analysis was performed at the Southwestern Indian Polytechnic Institute to make projections about melting permafrost and its impact on Alaskan native communities. These maps show the reduction in permafrost projected to the year 2050, and can help tribal governments with response and planning for future environmental change.

http://www.sipi.edu/

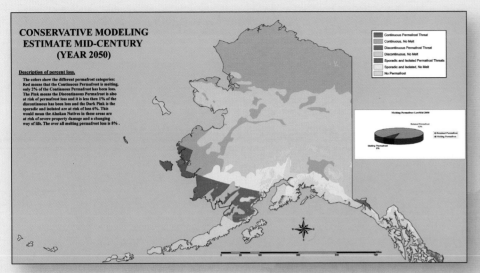

CONSERVATIVE MODELING ESTIMATE MID-CENTURY (YEAR 2050)

Description of percent loss.

The colors show the different Permafrost is melting. Red means that the Continuous Permafrost is melting, only 2% of the Continuous Permafrost has been loss. The Pink means the Discontinuous Permafrost is also at risk of permafrost loss and it is less then 1% of the discontinuous has been loss and the Dark Pink is the sporadic and isolated are at risk of loss 6%. This would mean the Alaskan Natives in these areas are at risk of severe property damage and a changing way of life. The over all melting permafrost loss is 8% .

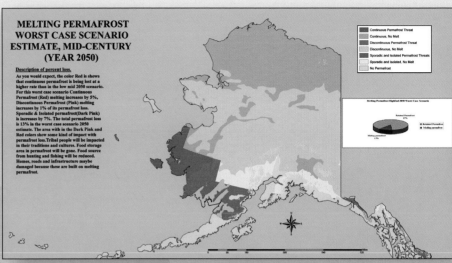

MELTING PERMAFROST WORST CASE SCENARIO ESTIMATE, MID-CENTURY (YEAR 2050)

Description of percent loss.

As you would expect, the color Red is shows that continuous permafrost is being lost at a higher rate than in the low mid 2050 scenario. For this worst case scenario Continuous Permafrost (Red) melting increases by 5%, Discontinuous Permafrost (Pink) melting increases by 1% of its permafrost loss. Sporadic & Isolated permafrost(Dark Pink) is increases by 13% in the worst case scenario 2050 estimate. The total permafrost loss is 13% in the worst case scenario 2050 estimate. The area with in the Dark Pink and Red colors show some kind of impact with permafrost loss.Tribal people will be impacted in their traditions and cultures. Food storage area in permafrost will be gone. Food source from hunting and fishing will be reduced. Homes, roads and infrastructure maybe damaged because these are built on melting permafrost.

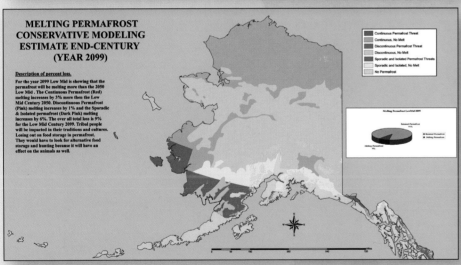

MELTING PERMAFROST CONSERVATIVE MODELING ESTIMATE END-CENTURY (YEAR 2099)

Description of percent loss.

For the year 2099 Low Mid is showing that the permafrost will be melting more than the 2050 Low Mid . The Continuous Permafrost (Red) melting increases by 3% more then the Low Mid Century 2050. Discontinuous Permafrost (Pink) melting increases by 1% and the Sporadic & Isolated permafrost (Dark Pink) melting increases by 6%. The over all total loss is 9% for the Low Mid Century 2099. Tribal people will be impacted in their traditions and cultures. Losing out on food storage in permafrost. They would have to look for alternative food storage and hunting because it will have an effect on the animals as well.

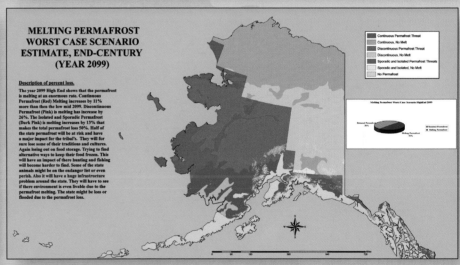

MELTING PERMAFROST WORST CASE SCENARIO ESTIMATE, END-CENTURY (YEAR 2099)

Description of percent loss.

The year 2099 High End shows that the permafrost is melting at an enormous rate. Continuous Permafrost (Red) Melting increases by 11% more than then the low mid 2099. Discontinuous Permafrost (Pink) is melting has increase by 26%. The Isolated and Sporadic Permafrost (Dark Pink) is melting increases by 13% that makes the total permafrost loss 50%. Half of the state permafrost will be at risk and have a major impact for the tribal's. They will for sure lose some of their traditions and cultures. Again losing out on food storage. Trying to find alternative ways to keep their food frozen. This will have an impact of there hunting and fishing will become harder to find. Some of the state animals might be on the endanger list or even perish. Also it will have a huge infrastructure problem around the state. They will have to see if there environent is even livable due to the permafrost melting. The state might be loss or flooded due to the permafrost loss.

GLORIA Mapping Program

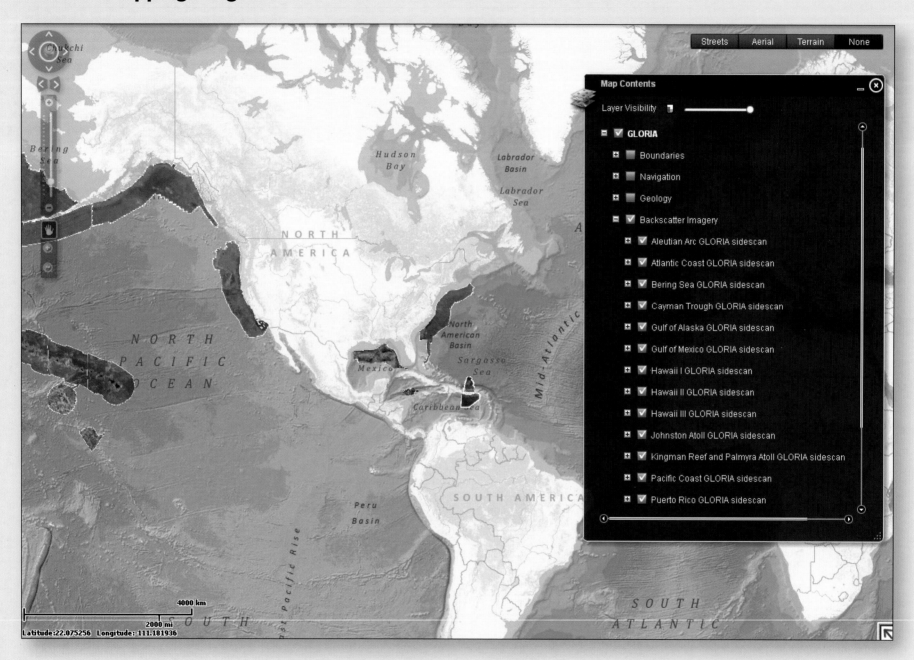

What defines the extended continental shelf of the United States and what resources exist there? To answer these questions, in 1984 a plan was developed specifically to map the deep ocean out to the 200-nautical-mile limit of the US Exclusive Economic Zone. The outcome was unprecedented, with images of the sea floor that continue to be examined more than twenty years after the program ended.

http://coastalmap.marine.usgs.gov/gloria/

Known and Potential Streams at Morristown National Historical Park

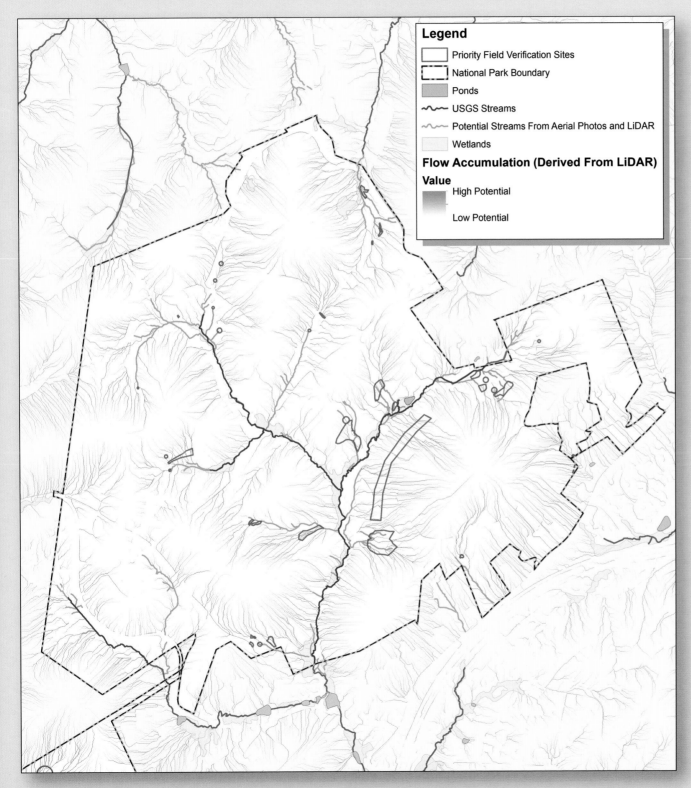

Legend

- ☐ Priority Field Verification Sites
- ☐ National Park Boundary
- ▨ Ponds
- ∿ USGS Streams
- ∿ Potential Streams From Aerial Photos and LiDAR
- ☐ Wetlands

Flow Accumulation (Derived From LiDAR)
Value

High Potential

Low Potential

The National Park Service sought a better understanding of historical events that occurred in Morristown National Historical Park. The Continental Army chose to overwinter there from 1777 to 1780 because of the distinctive topographic setting and useful local resources, including the many streams. However, current data showed few streams on the large and heavily wooded site. Using lidar, flow direction surfaces and flow accumulation lines were calculated to identify likely additional streams and wetlands. This allowed the field crew to prioritize searches on the ground, saving time, effort, and money.

http://www.nps.gov/morr/
naturescience/index.htm

US DEPARTMENT OF STATE

Ethiopia: Three Ways of Looking at HIV Distribution

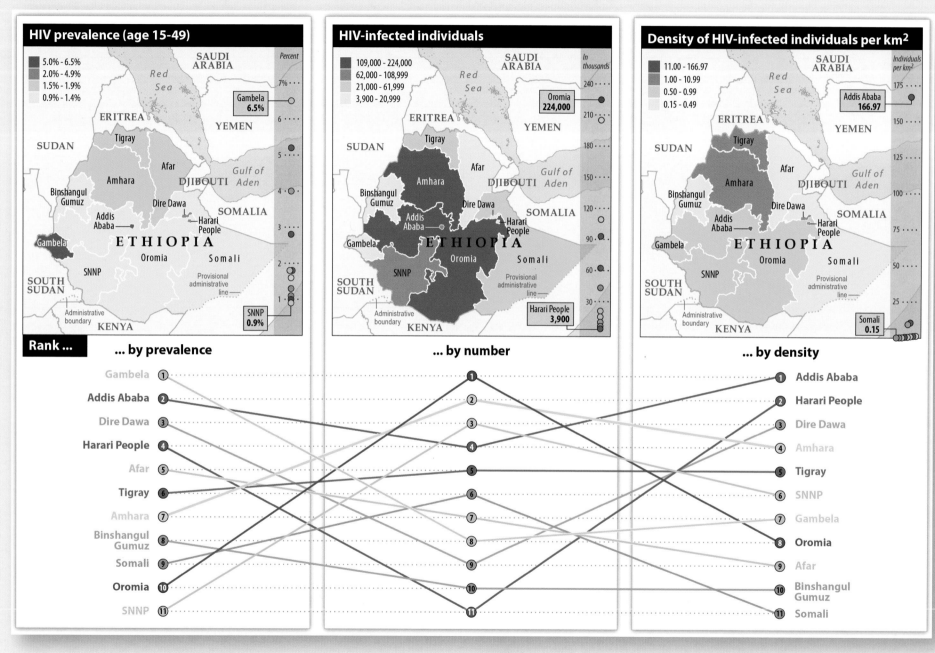

Ethiopia is among the countries most affected by the HIV epidemic. According to a 2012 report by the Ethiopian government, the country has approximately 800,000 people living with HIV and about one million AIDS orphans. These maps show HIV distribution in Ethiopia based on prevalence, number infected, and density in 2011. This information can help with family planning initiatives and education outreach.

http://www.state.gov/s/inr/hiu/

Pakistan: Humanitarian Crises in 2011

During regional conflicts, displaced people are at great risk and in need of humanitarian assistance. GIS plays a key role in creating maps that show how displaced people are moving and how many are involved. Maps such as this provide excellent summaries of complex movements and help improve the assistance process.

http://www.state.gov/s/inr/hiu/

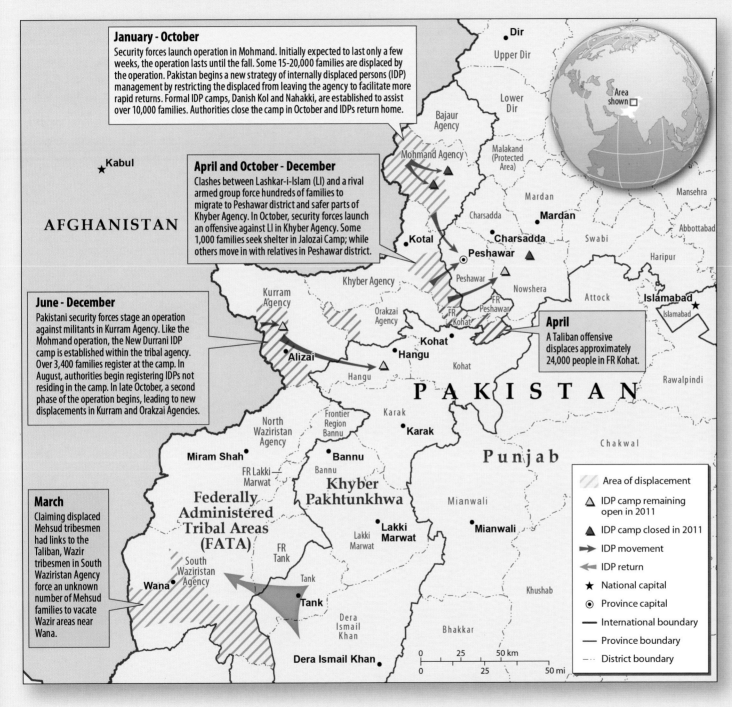

January - October

Security forces launch operation in Mohmand. Initially expected to last only a few weeks, the operation lasts until the fall. Some 15-20,000 families are displaced by the operation. Pakistan begins a new strategy of internally displaced persons (IDP) management by restricting the displaced from leaving the agency to facilitate more rapid returns. Formal IDP camps, Danish Kol and Nahakki, are established to assist over 10,000 families. Authorities close the camp in October and IDPs return home.

April and October - December

Clashes between Lashkar-i-Islam (LI) and a rival armed group force hundreds of families to migrate to Peshawar district and safer parts of Khyber Agency. In October, security forces launch an offensive against LI in Khyber Agency. Some 1,000 families seek shelter in Jalozai Camp; while others move in with relatives in Peshawar district.

June - December

Pakistani security forces stage an operation against militants in Kurram Agency. Like the Mohmand operation, the New Durrani IDP camp is established within the tribal agency. Over 3,400 families register at the camp. In August, authorities begin registering IDPs not residing in the camp. In late October, a second phase of the operation begins, leading to new displacements in Kurram and Orakzai Agencies.

April

A Taliban offensive displaces approximately 24,000 people in FR Kohat.

March

Claiming displaced Mehsud tribesmen had links to the Taliban, Wazir tribesmen in South Waziristan Agency force an unknown number of Mehsud families to vacate Wazir areas near Wana.

Legend

- Area of displacement
- △ IDP camp remaining open in 2011
- ▲ IDP camp closed in 2011
- ➤ IDP movement
- ◄ IDP return
- ★ National capital
- ◉ Province capital
- ― International boundary
- ― Province boundary
- ⋯ District boundary

0 25 50 km
0 25 50 mi

Syria: 2012 Population Displacement in Review

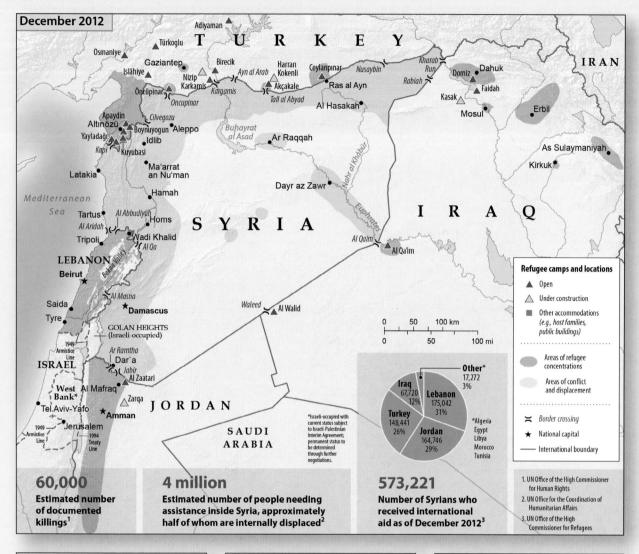

December 2012

An increasing number of Syrians have been displaced to Lebanon, Jordan, Turkey, and Iraq as a result of regional conflict, and are living in camps or other accommodations in these countries. This map shows the displacement situation in Syria and neighboring countries in 2012. These graphic products help inform aid organizations and workers in the management and distribution of resources to refugees.

http://www.state.gov/s/inr/hiu/

Refugee camps and locations

▲ Open
△ Under construction
■ Other accommodations (e.g., host families, public buildings)

⬤ Areas of refugee concentrations
Areas of conflict and displacement

⋈ Border crossing
★ National capital
— International boundary

Pie chart:
- Iraq 67,720 12%
- Lebanon 175,042 31%
- Turkey 148,441 26%
- Jordan 164,746 29%
- Other* 17,272 3%

*Algeria Egypt Libya Morocco Tunisia

*Israeli-occupied with current status subject to Israeli-Palestinian Interim Agreement; permanent status to be determined through further negotiations.

60,000
Estimated number of documented killings[1]

4 million
Estimated number of people needing assistance inside Syria, approximately half of whom are internally displaced[2]

573,221
Number of Syrians who received international aid as of December 2012[3]

1. UN Office of the High Commissioner for Human Rights
2. UN Office for the Coordination of Humanitarian Affairs
3. UN Office of the High Commissioner for Refugees

January

April

September

World Humanitarian Day 2012: Aid Workers in Harm's Way

Since 2001, the number of aid worker victims who have been killed, wounded/injured, or kidnapped in security incidents has risen dramatically. The majority of these victims are national staff aid workers. Aid workers can be the targets of combatants or collateral casualties caught in the crossfire. The information provided in this map can help aid organizations with security management and safety.

http://www.state.gov/s/inr/hiu/

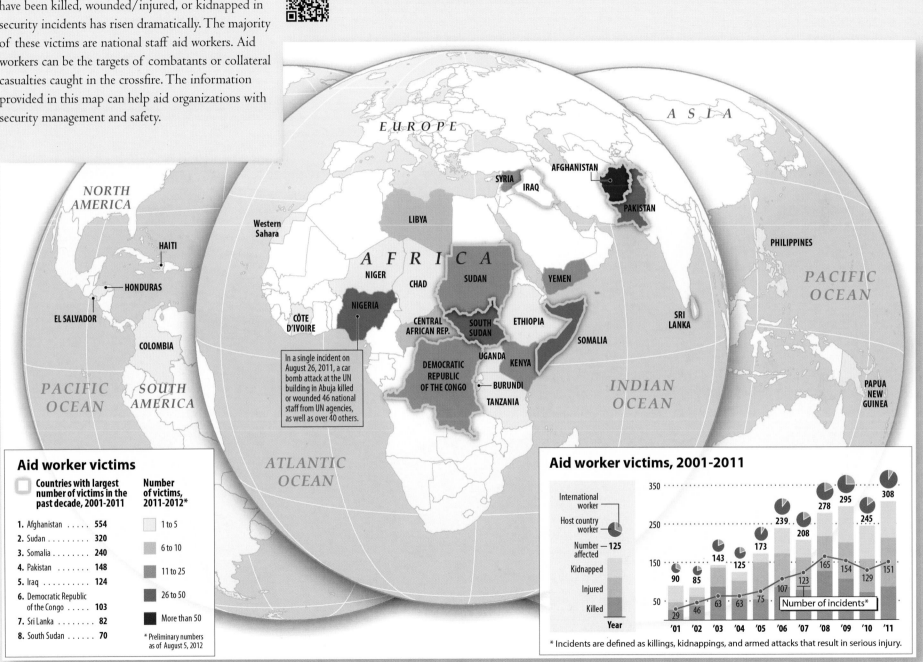

In a single incident on August 26, 2011, a car bomb attack at the UN building in Abuja killed or wounded 46 national staff from UN agencies, as well as over 40 others.

Aid worker victims

Countries with largest number of victims in the past decade, 2001-2011

1. Afghanistan 554
2. Sudan 320
3. Somalia 240
4. Pakistan 148
5. Iraq 124
6. Democratic Republic of the Congo 103
7. Sri Lanka 82
8. South Sudan 70

Number of victims, 2011-2012*

- 1 to 5
- 6 to 10
- 11 to 25
- 26 to 50
- More than 50

* Preliminary numbers as of August 5, 2012

Aid worker victims, 2001-2011

International worker
Host country worker

Number affected — 125
Kidnapped
Injured
Killed

Year

Year	Number affected	Number of incidents*
'01	90	29
'02	85	46
'03	143	63
'04	125	63
'05	173	75
'06	239	107
'07	208	123
'08	278	165
'09	295	154
'10	245	129
'11	308	151

* Incidents are defined as killings, kidnappings, and armed attacks that result in serious injury.

Maritime Claim Lines and Limits near the Strait of Hormuz

The narrow Strait of Hormuz between the Persian Gulf and the Arabian Sea is the transit passage for a significant portion of the world's oil and gas. Part of this international strait passes through the 12-nautical-mile territorial sea claims of Iran and Oman. This map shows the claimed territorial seas and negotiated continental shelf claims near the Strait of Hormuz. Information such as this is important for diplomatic relationships and decisions.

 https://hiu.state.gov/data/data.aspx

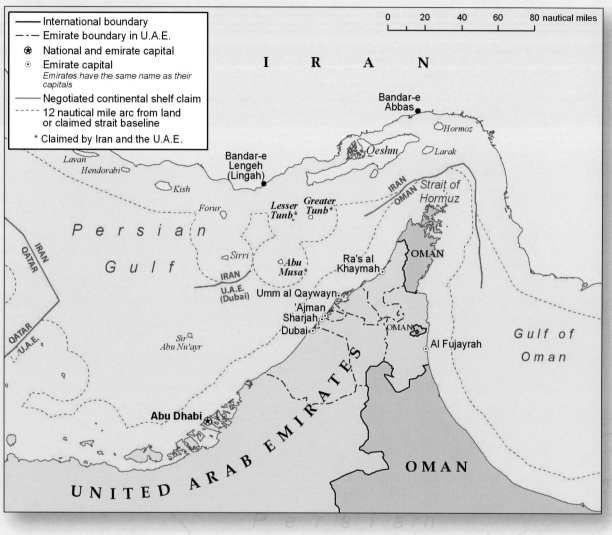

Legend:
— International boundary
–·– Emirate boundary in U.A.E.
⊗ National and emirate capital
⊙ Emirate capital
  Emirates have the same name as their capitals
— Negotiated continental shelf claim
---- 12 nautical mile arc from land or claimed strait baseline
  * Claimed by Iran and the U.A.E.

0   20   40   60   80 nautical miles

Sea Ice Extent and Arctic Nation Limits

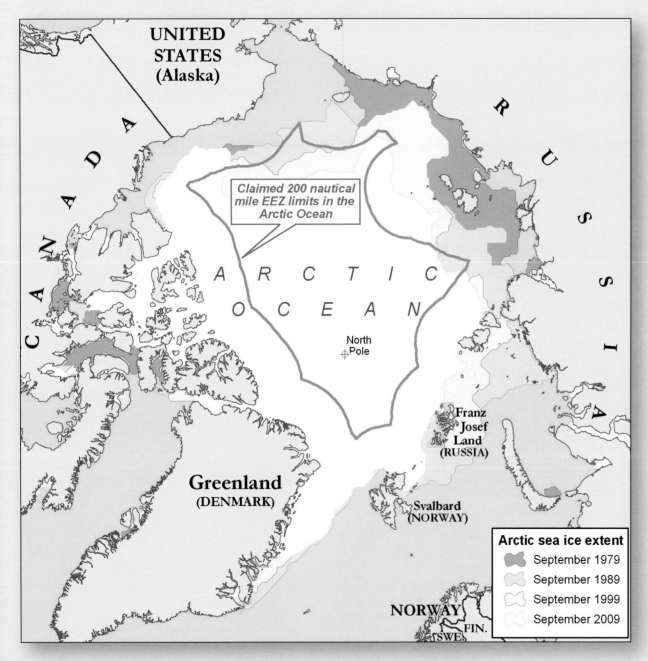

UNITED STATES (Alaska)

R U S S I A

C A N A D A

Claimed 200 nautical mile EEZ limits in the Arctic Ocean

A R C T I C

O C E A N

North Pole

Franz Josef Land (RUSSIA)

Greenland (DENMARK)

Svalbard (NORWAY)

NORWAY

FIN.

SWE.

Arctic sea ice extent

	September 1979
	September 1989
	September 1999
	September 2009

With the acceleration of Arctic ice melt over the last few decades, extracting resources in the Arctic Ocean becomes more viable. Ice-free areas are opening up, which enables nations to fish in waters previously difficult to exploit. This map illustrates areas that are opening up to outside exploitation, and are critical to guiding official decisions for ocean planning.

https://hiu.state.gov/data/data.aspx

Syria: Halab (Aleppo) Governorate

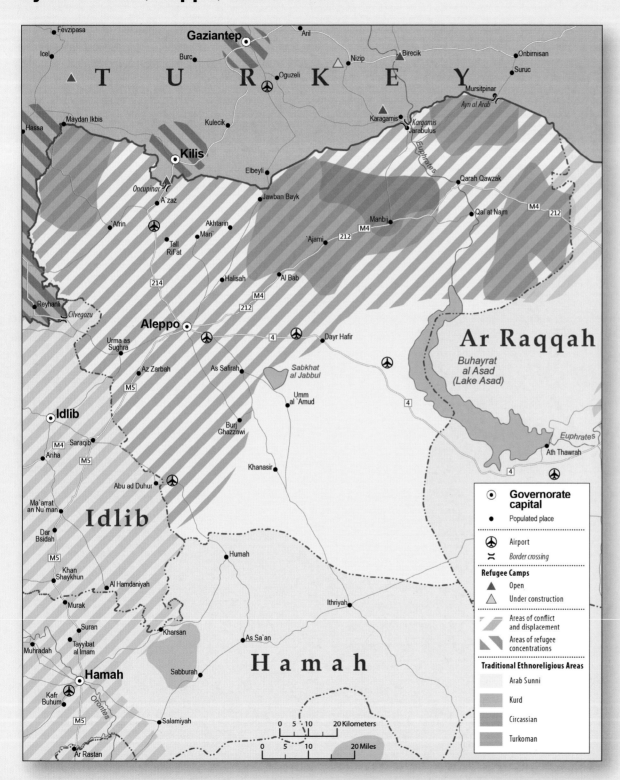

With the growing crisis in Syria, a series of maps was created to help decision makers better understand the complex situation occurring in seven key Syrian governorates. Prior to the production of this series, officials relied on maps from a variety of sources. This map in the series takes numerous sets of information and combines them into a single manageable product for decision makers to use more easily.

https://hiu.state.gov/data/data.aspx

Legend:

- ⊙ **Governorate capital**
- • Populated place
- ✈ Airport
- ⋈ Border crossing

Refugee Camps
- ▲ Open
- △ Under construction

- Areas of conflict and displacement
- Areas of refugee concentrations

Traditional Ethnoreligious Areas
- Arab Sunni
- Kurd
- Circassian
- Turkoman

0 5 10 20 Kilometers
0 5 10 20 Miles

January 18, 1974 Egyptian-Israeli Disengagement Agreement (Sinai I)

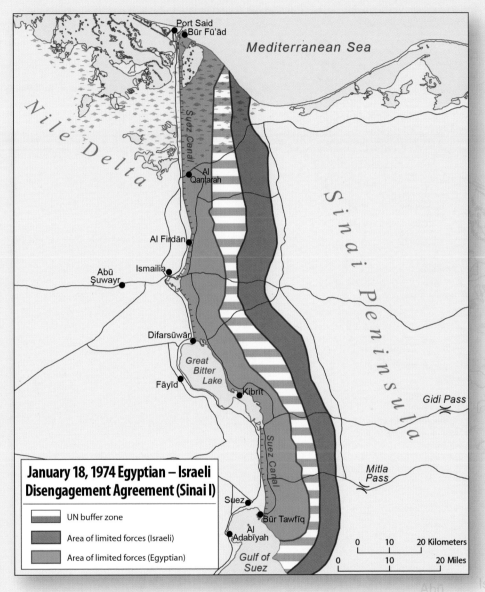

January 18, 1974 Egyptian – Israeli Disengagement Agreement (Sinai I)

UN buffer zone

Area of limited forces (Israeli)

Area of limited forces (Egyptian)

At the request of the State Department Office of the Historian, a series of historic maps was created to cover the Arab-Israeli Disputes of 1974–1976. Since no GIS data existed for these boundaries, the treaty and disengagement lines were created from original sources. The displayed map illustrates the initial disengagement agreement between Egypt and Israel.

https://hiu.state.gov/data/data.aspx

Provinces and Districts of Afghanistan, February 2012

A Jabal us Sarāj
B Sayyid Khayl
C Hişah-e Awal-e Kōhistān
D Hişah-e Duwum-e Kōhistān
E Mīr Bachah Kōt
F Kalakān
G Walī Muḥammad Shahīd Khūgyānī

Up-to-date and accurate administrative boundaries are necessary for US Department of State operations. This map of Afghanistan shows the districts and province centers across the entire country. Keeping officials informed about the regional makeup of other countries is important for cultural and administrative purposes.

https://hiu.state.gov/data/data.aspx

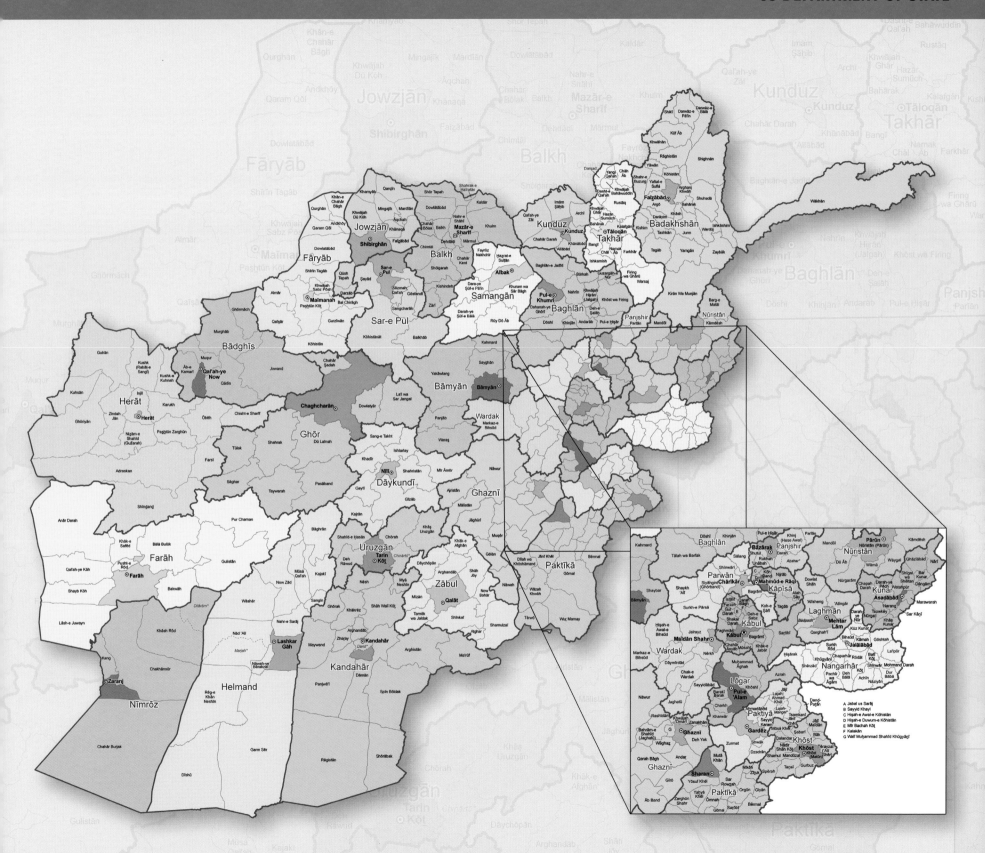

A Jabal us Sarāj
B Sayyid Khayl
C Hisah-e Awal-e Kōhistān
D Hisah-e Duwum-e Kōhistān
E Mīr Bachah Kōt
F Kalakān
G Walī Muhammad Shahīd Khūgyāgī

US DEPARTMENT OF TRANSPORTATION

Federal Lands Highway Mapping Application

The primary purpose of the Office of Federal Lands Highway is to provide resources for public roads that service the transportation needs of federal and Indian lands. This GIS web application shows the map and route location tool for the National Park Service. This tool helps in monitoring road conditions and in planning for infrastructure investment.

http://flh.fhwa.dot.gov/

Structurally Deficient Bridges

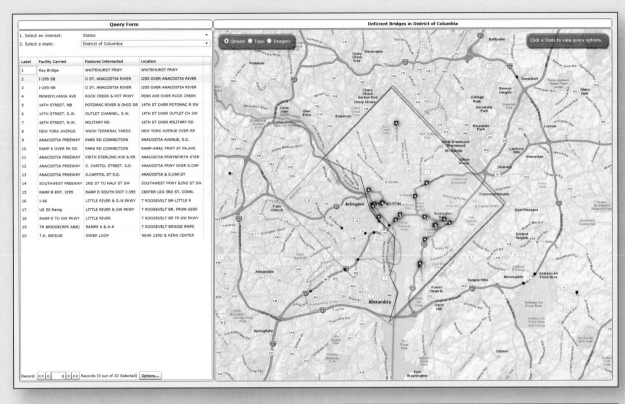

The US Department of Transportation informs Congress about the condition of bridges across the United States and identifies bridges categorized as deficient. This web application provides information about structurally deficient bridges on the national highway system. Members of Congress as well as the public can obtain a list of deficient bridges in their area, and plan accordingly.

 http://flh.fhwa.dot.gov/

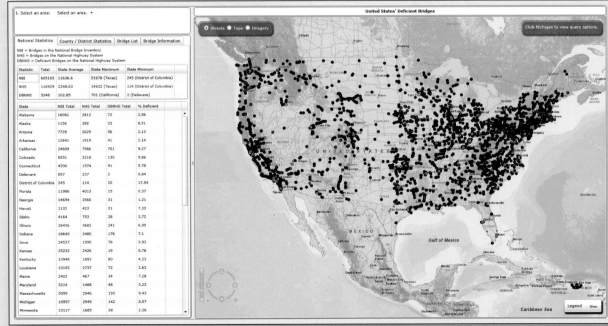

Draft Electronic Airport Layout Plan for Valley International Airport

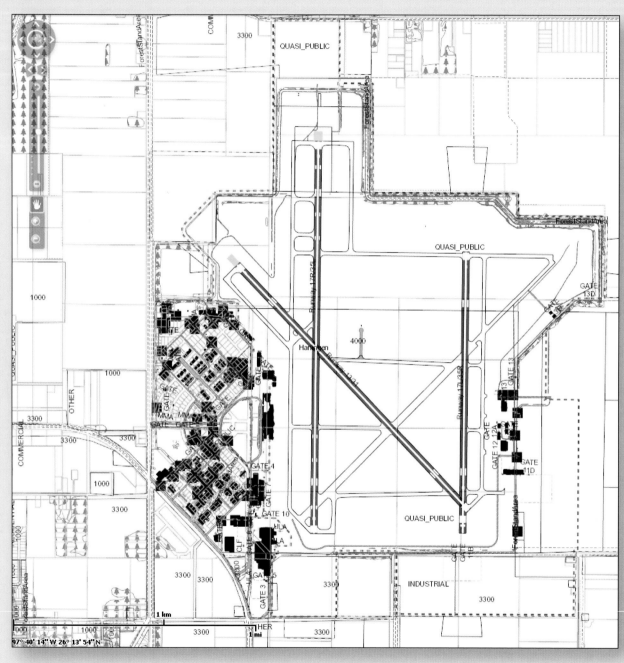

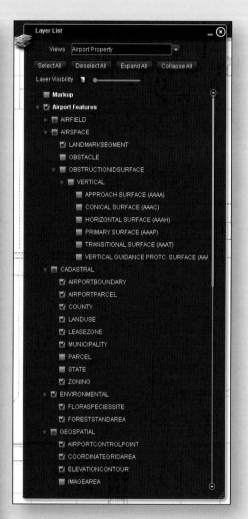

http://www.faa.gov/
airports/planning_capacity/
airports_gis_electronic_alp/

The Federal Aviation Administration (FAA) needs to meet future air traffic demands and minimize airport congestion. To achieve these goals, airport and aeronautical data must be collected and compiled. This image of Valley International Airport in Harlingen, Texas, represents early work by the FAA to develop a system that will support the airports GIS program and thus keep air traffic flowing smoothly.

Washington, DC, Low Sector Airspace

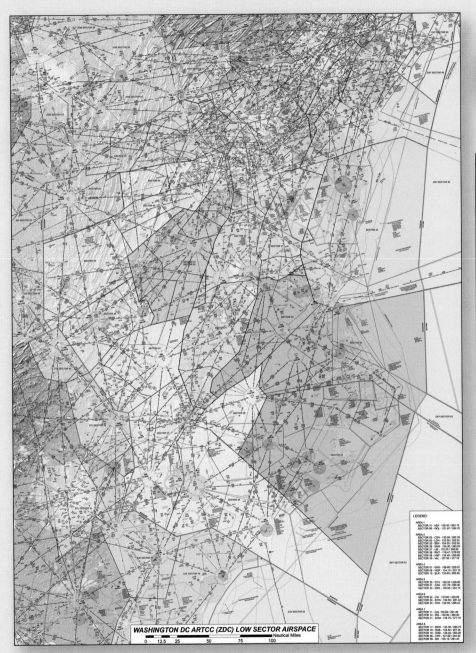

PROJECT BENEFITS

As a result of using ArcGIS in air traffic control centers, the Federal Aviation Administration was able to save over $250,000.

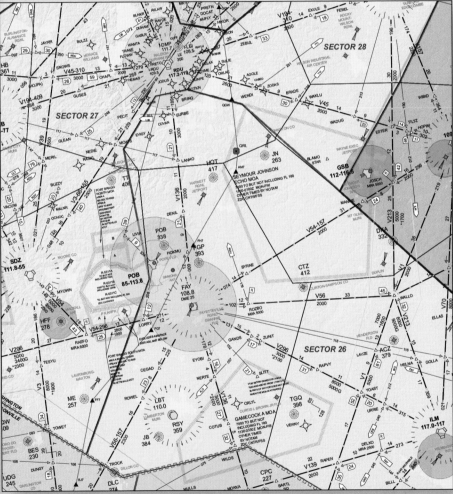

The FAA wants to modernize its mapping system and integrate existing tools to produce better maps. The system is being updated to ArcGIS across twenty-one Air Route Traffic Control Centers. This new system will allow the Administration to produce high-quality air maps and data. This map shows a Washington, DC, control center.

http://www.faa.gov/about/office_org/headquarters_offices/ato/artcc/

US DEPARTMENT OF VETERANS AFFAIRS

Mapping the 2011–2012 Influenza Season by VA facility

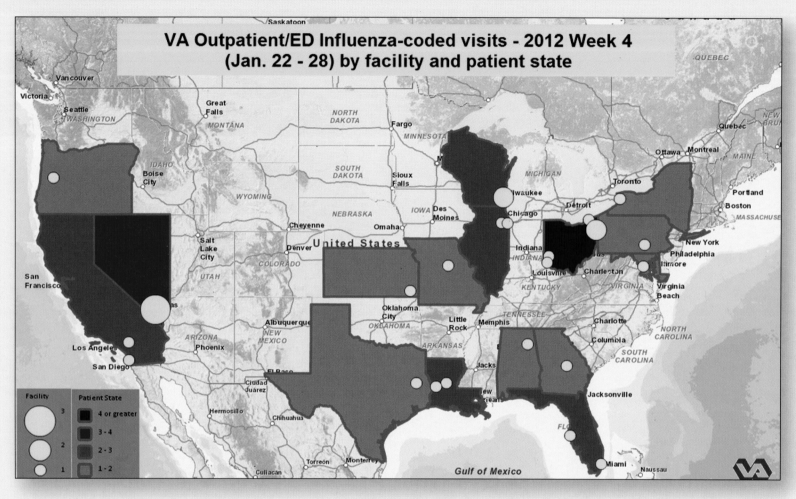

VA Outpatient/ED Influenza-coded visits - 2012 Week 4
(Jan. 22 - 28) by facility and patient state

Flu outbreaks are a serious matter, and it's important to understand trends and patterns of the past to prepare for future events. This map uses GIS to show the number of outpatient and emergency department visits with an influenza diagnosis. Better planning of preventive measures contributes to minimizing the occurrence of outbreaks in the future.

http://www.publichealth.va.gov/

VHA Health Care Supply and Demand

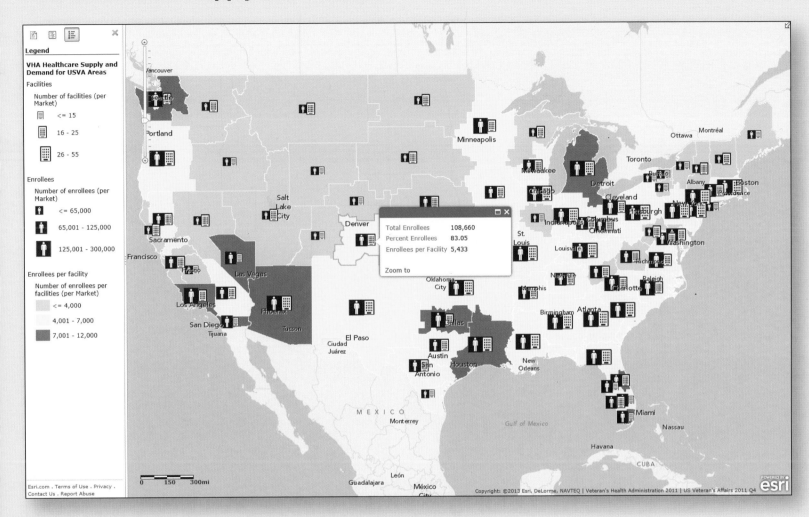

In an effort to better understand the location of care sites compared to the demand for care, the Veterans Health Administration created a GIS-driven online map of facilities and the number of people each was required to support. This map shows the demand being placed on each facility and helps the agency plan against potentially overwhelming a facility or underserving a community.

http://www.va.gov/health/default.asp

Health Care Safety Measures for the Veterans Health Administration

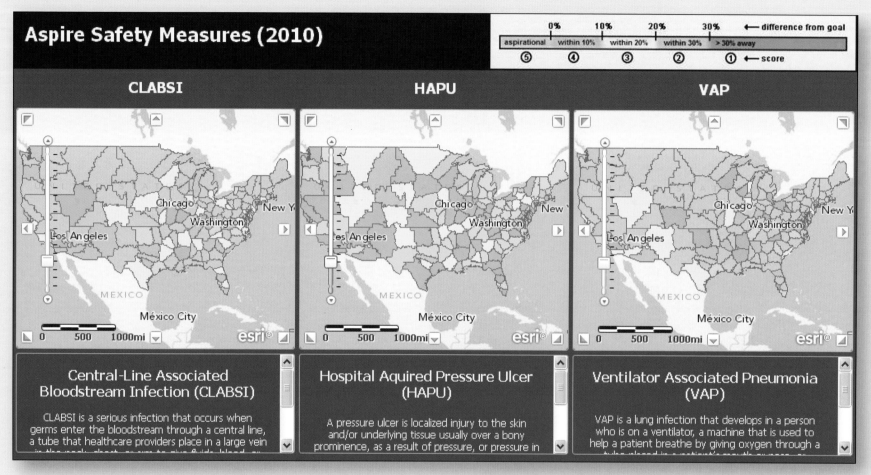

Aspire Safety Measures (2010)

0%   10%   20%   30%   ← difference from goal
aspirational | within 10% | within 20% | within 30% | > 30% away
⑤   ④   ③   ②   ① ← score

CLABSI

HAPU

VAP

Central-Line Associated Bloodstream Infection (CLABSI)

CLABSI is a serious infection that occurs when germs enter the bloodstream through a central line, a tube that healthcare providers place in a large vein

Hospital Aquired Pressure Ulcer (HAPU)

A pressure ulcer is localized injury to the skin and/or underlying tissue usually over a bony prominence, as a result of pressure, or pressure in

Ventilator Associated Pneumonia (VAP)

VAP is a lung infection that develops in a person who is on a ventilator, a machine that is used to help a patient breathe by giving oxygen through a

The Department of Veterans Affairs (VA) needed to determine improvement opportunities for all of their hospitals. Using ArcGIS.com, the VA developed an interactive dashboard that allows users to compare various safety values and scores, documenting quality and safety in comparison to goals. This tool provides compelling pictures of performance that help the VA meet its objectives.

http://www.va.gov/health/default.asp

VHA Health Care Service and Live Weather Events

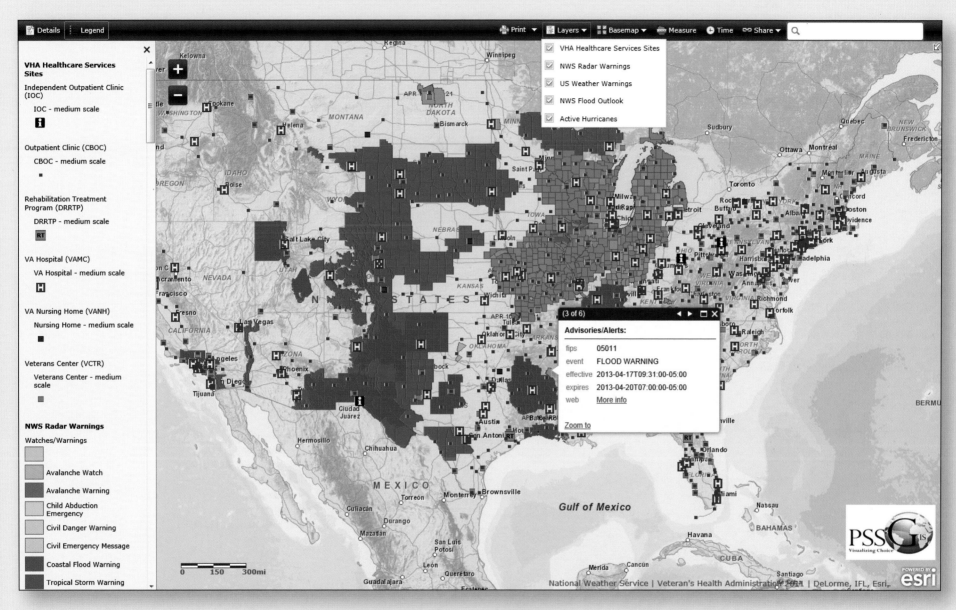

Because major weather events can cause damage and disrupt health services, information about weather impacts on facilities needs to be disseminated quickly and accurately to plan for a response. This web map uses GIS to show weather data and Department of Veterans Affairs (VA) health care sites, and can be viewed by VA leadership or the public on a tablet or smartphone. This service allows the VA to provide up-to-date information and keep decision makers and the public informed.

http://www.va.gov/health/default.asp

Improving Veterans' Access to Services

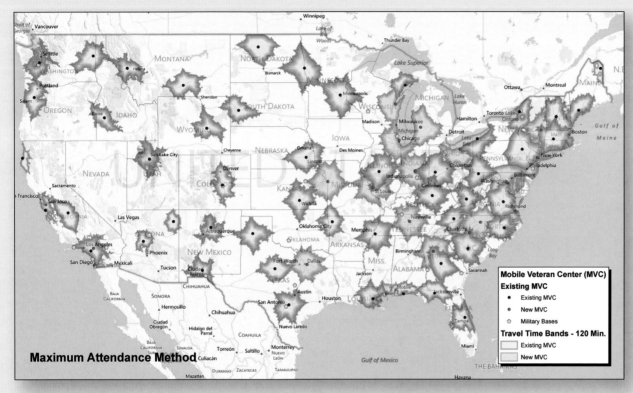

Maximum Attendance Method

Many veterans transitioning from military to civilian life are not within reasonable driving distance of a Vet Center. The accompanying maps show how the Department of Veterans Affairs used GIS to help plan the location and use of Mobile Vet Centers (MVC) to improve care access for these vets. GIS and geographical analysis provide a direct benefit to seeing that our servicemen and women receive care.

 http://www.va.gov/health/default.asp

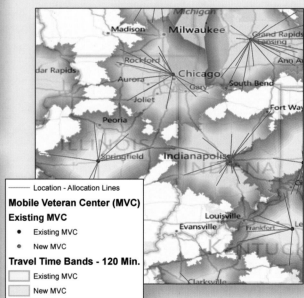

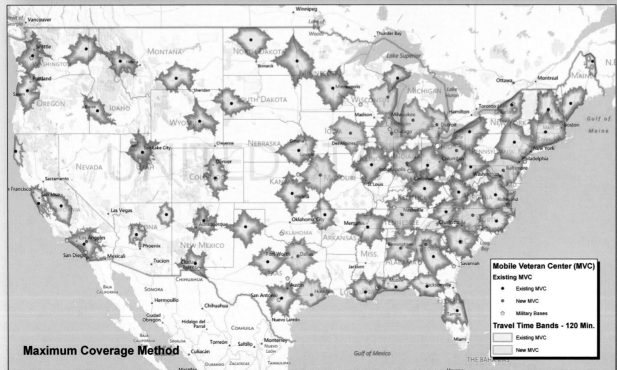

Maximum Coverage Method

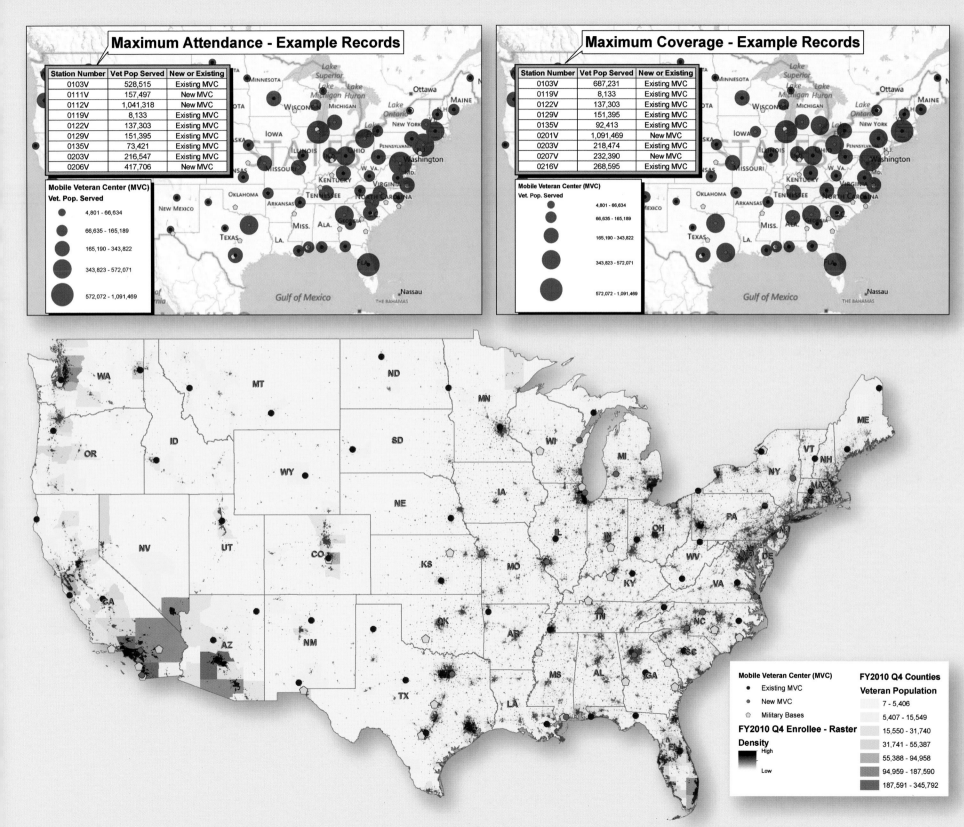

Maximum Attendance - Example Records

Station Number	Vet Pop Served	New or Existing
0103V	528,515	Existing MVC
0111V	157,497	New MVC
0112V	1,041,318	New MVC
0119V	8,133	Existing MVC
0122V	137,303	Existing MVC
0129V	151,395	Existing MVC
0135V	73,421	Existing MVC
0203V	216,547	Existing MVC
0206V	417,706	New MVC

Mobile Veteran Center (MVC)
Vet. Pop. Served
- 4,801 - 66,634
- 66,635 - 165,189
- 165,190 - 343,822
- 343,823 - 572,071
- 572,072 - 1,091,469

Maximum Coverage - Example Records

Station Number	Vet Pop Served	New or Existing
0103V	687,231	Existing MVC
0119V	8,133	Existing MVC
0122V	137,303	Existing MVC
0129V	151,395	Existing MVC
0135V	92,413	Existing MVC
0201V	1,091,469	New MVC
0203V	218,474	Existing MVC
0207V	232,390	New MVC
0216V	268,595	Existing MVC

Mobile Veteran Center (MVC)
Vet. Pop. Served
- 4,801 - 66,634
- 66,635 - 165,189
- 165,190 - 343,822
- 343,823 - 572,071
- 572,072 - 1,091,469

Mobile Veteran Center (MVC)
- Existing MVC
- New MVC
- Military Bases

FY2010 Q4 Enrollee - Raster Density
- High
- Low

FY2010 Q4 Counties
Veteran Population
- 7 - 5,406
- 5,407 - 15,549
- 15,550 - 31,740
- 31,741 - 55,387
- 55,388 - 94,958
- 94,959 - 187,590
- 187,591 - 345,792

VHA Medicaid Expansion and Health Insurance Marketplace Web App

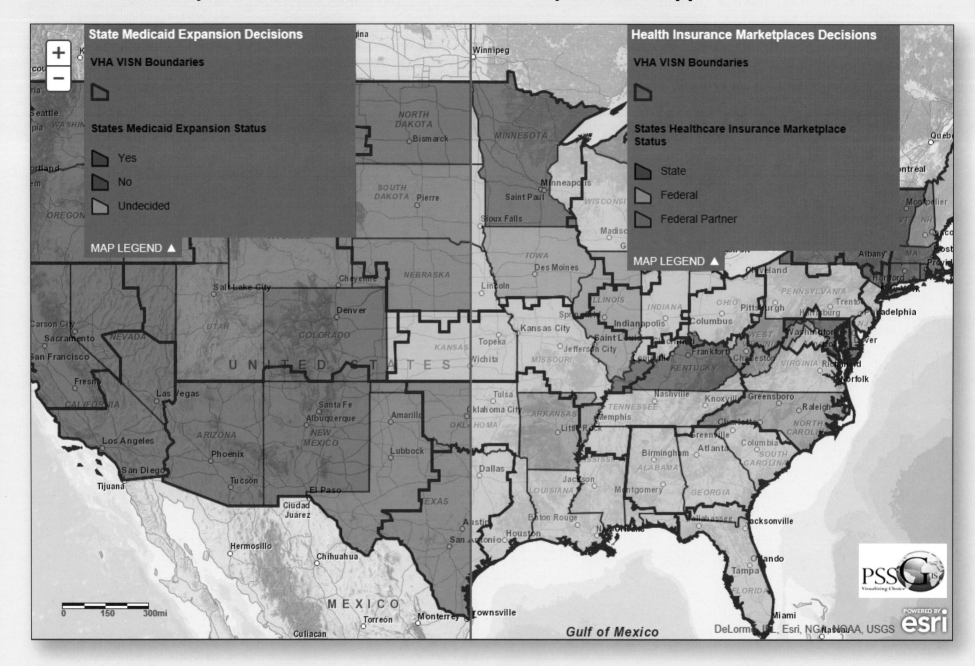

Veterans Health Administration leadership needs to visualize the potential relationship between state Medicaid expansion decisions and health insurance marketplace decisions. This web GIS application was developed to allow comparison of the data and to foster a better understanding of the status of health coverage in America. This particular map was current as of April 15, 2013.

http://www.va.gov/health/default.asp

Veterans Health Administration GIS Planning Portal

Planners for the Department of Veterans Affairs (VA) were looking for a way to develop better solutions that would improve medical access for their patients. The Veterans Health Administration GIS Planning Portal was developed to provide support for VA planners to determine optimal site locations and quantify patient access statistics. The analysis capabilities of this online solution allow the VA to plan effectively and efficiently for the future.

http://www.va.gov/health/default.asp

INDEPENDENT FEDERAL AGENCIES

US General Services Administration Maps

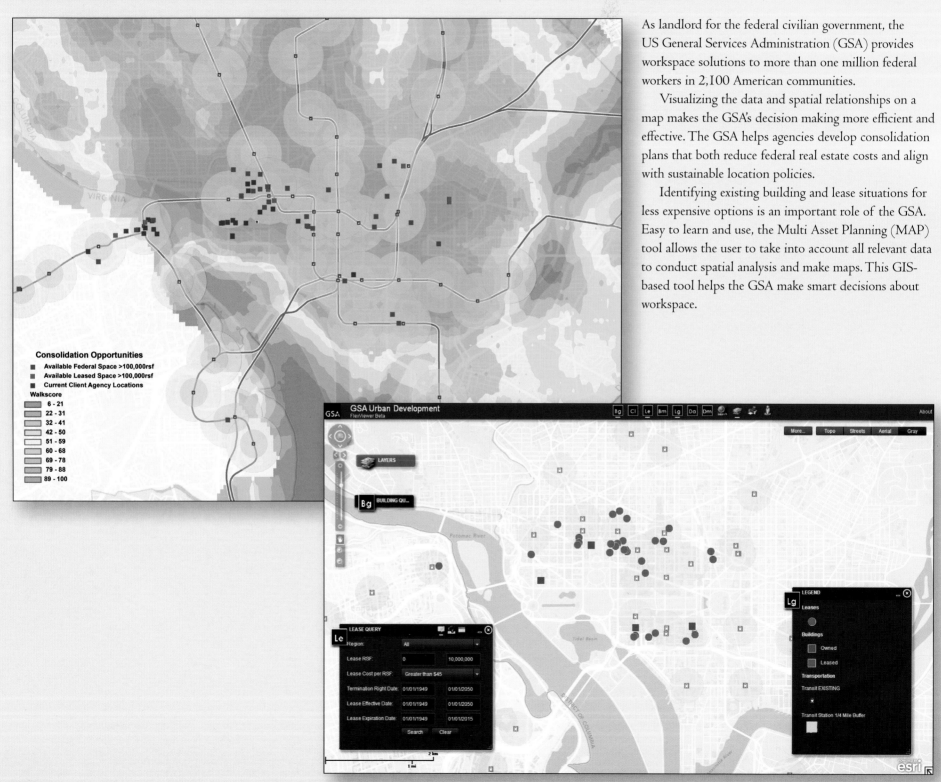

Consolidation Opportunities

- Available Federal Space >100,000rsf
- Available Leased Space >100,000rsf
- Current Client Agency Locations

Walkscore

- 6 - 21
- 22 - 31
- 32 - 41
- 42 - 50
- 51 - 59
- 60 - 68
- 69 - 78
- 79 - 88
- 89 - 100

As landlord for the federal civilian government, the US General Services Administration (GSA) provides workspace solutions to more than one million federal workers in 2,100 American communities.

Visualizing the data and spatial relationships on a map makes the GSA's decision making more efficient and effective. The GSA helps agencies develop consolidation plans that both reduce federal real estate costs and align with sustainable location policies.

Identifying existing building and lease situations for less expensive options is an important role of the GSA. Easy to learn and use, the Multi Asset Planning (MAP) tool allows the user to take into account all relevant data to conduct spatial analysis and make maps. This GIS-based tool helps the GSA make smart decisions about workspace.

The public wants to know that funds are used wisely. These web maps keep Americans informed about where GSA is spending its funds from the American Recovery and Reinvestment Act. They show how the money was used to convert existing federal buildings into more environmentally friendly ones and to build new, energy-efficient structures. This online solution improves public awareness, and helps the GSA be more transparent.

PROJECT BENEFITS

GSA works with its customer agencies to make smart real estate decisions that both minimize costs and align with federal sustainability goals. Using ArcGIS, account managers can quickly compare agencies' current locations with available vacant space in nearby federally owned and leased buildings. This comprehensive awareness can promote good decision making and help government employees do their jobs better.

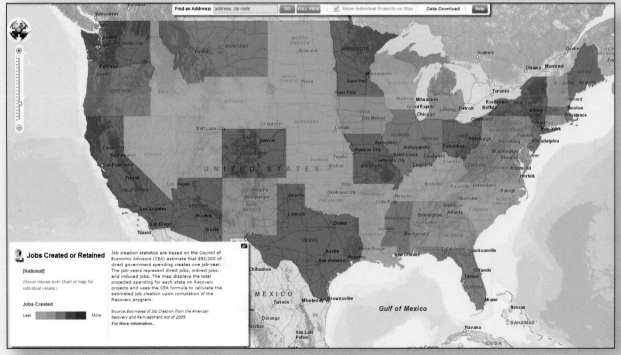

US General Services Administration Maps continued

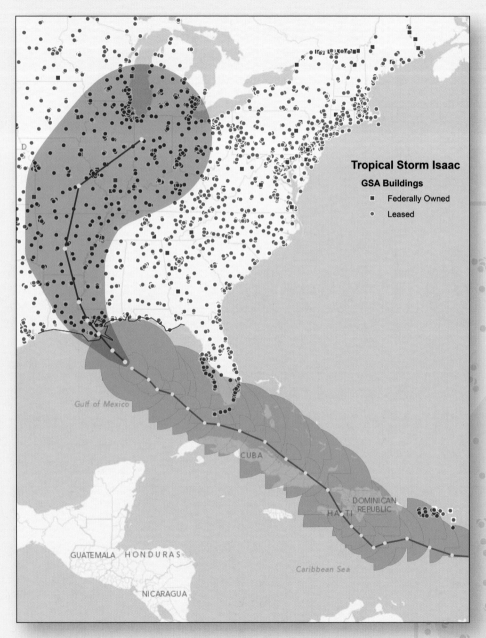

Tropical Storm Isaac

GSA Buildings

■ Federally Owned

● Leased

As Tropical Storm Isaac approached in 2012, federal employees and buildings lay in its path. The GSA's Office of Emergency Response and Recovery used ArcGIS to quickly determine risks, and alerted building management and federal employees to prepare. Coordination efforts were initiated with other federal agencies. Valuable time in planning for Isaac and the actions taken in advance of the storm helped reduce the cleanup costs.

 http://www.gsa.gov/portal/content/104444

Tegucigalpa, Honduras Bus Routes

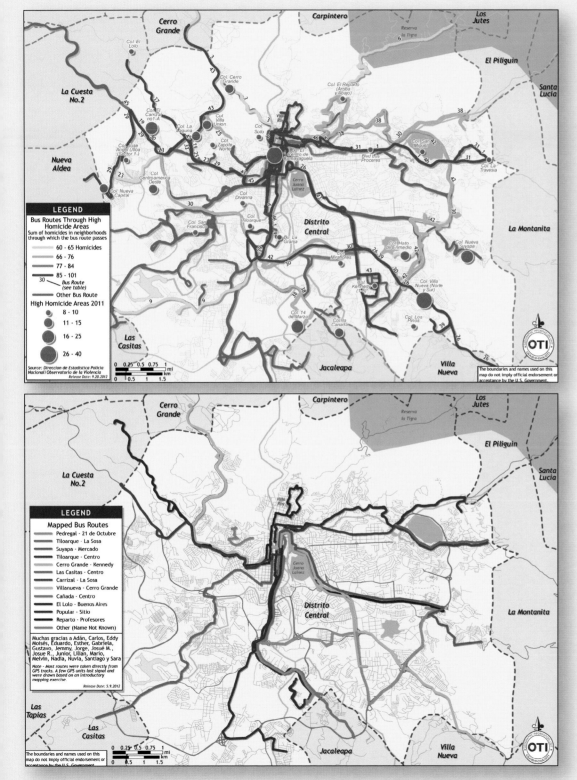

In Tegucigalpa, Honduras, buses are a very popular form of transportation. However, there is often crime on buses and bus route data is not available. The US Agency for International Development (USAID) was interested in learning more about where the buses go in order to plan citizen security projects. To accomplish this, USAID staff trained Honduran students on using GPS units and mapping technology to create the bus route map. The resulting bus route data was overlaid with existing crime data to suggest routes that might be more dangerous or would be good candidates for citizen security projects.

http://transition.usaid.gov/hn/

Every Door Direct Mail Online Application

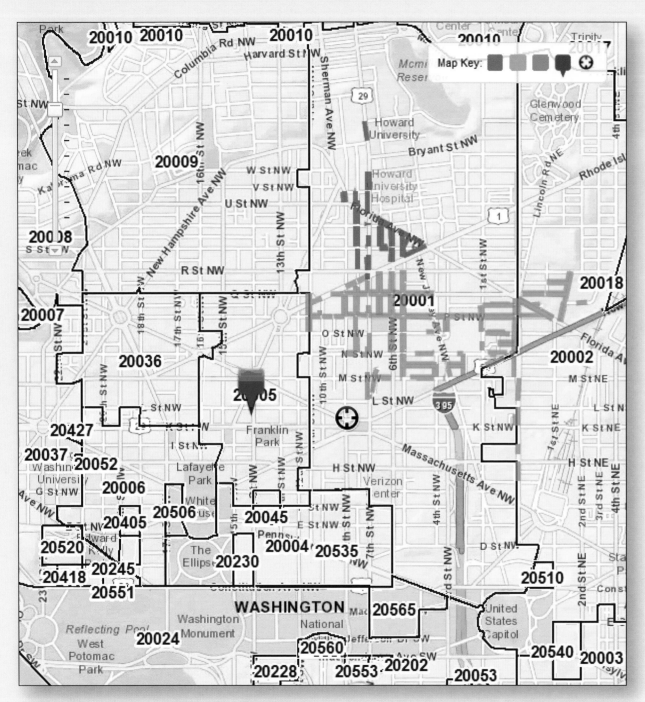

Direct mail is an effective way for businesses to get their message to consumers. Every Door Direct Mail from the US Postal Service enables direct mail customers to map out a target area, select a delivery route and mailing drop-off date, and pay online from their computers. Efficiencies using GIS help to keep distribution costs of advertising and other printed messaging low.

https://www.usps.com/business/every-door-direct-mail.htm

USAID GeoCenter Maps

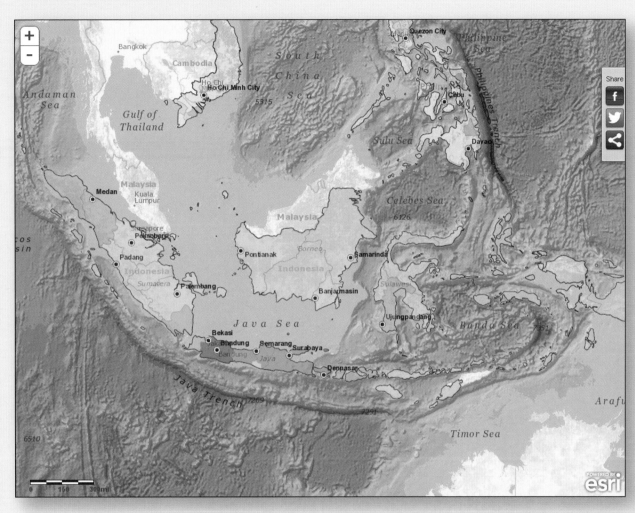

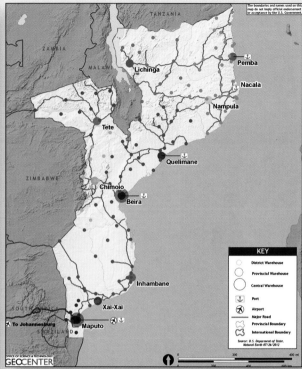

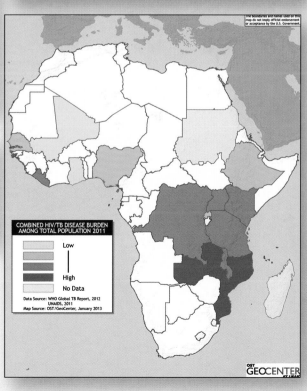

USAID officials need to understand local economies, health care, and disease so they can strategically allocate resources to assist counties and regions. The Center for the Application of Geospatial Analysis for Development (GeoCenter) uses GIS for strategic planning, monitoring and evaluating projects, and communicating results. These maps demonstrate how GIS helps officials meet the health needs of the populations they serve in Africa.

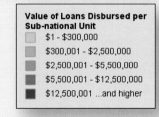

Value of Loans Disbursed per Sub-national Unit
- $1 - $300,000
- $300,001 - $2,500,000
- $2,500,001 - $5,500,000
- $5,500,001 - $12,500,000
- $12,500,001 ...and higher

http://transition.usaid.gov/scitech/gis.html

Navajo Nation Maps

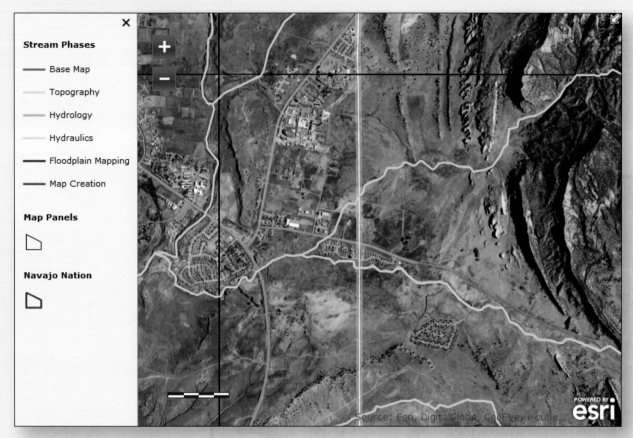

The Navajo Housing Authority (NHA) floodplain study project is a large-scale effort that requires the distribution of important project information, sharing of documents, and monitoring of project status. The website for this study is interactive and is a central location for stakeholders and project team members to view project updates as necessary.

The Navajo Housing Authority wants to minimize flood risk to existing homes and infrastructure. More than 3,000 miles of floodplains have been mapped for the housing authority. NHA uses these floodplain maps to determine suitable locations for new homes. This image provides a 3D perspective of the flood hazards in Kayenta, Arizona.

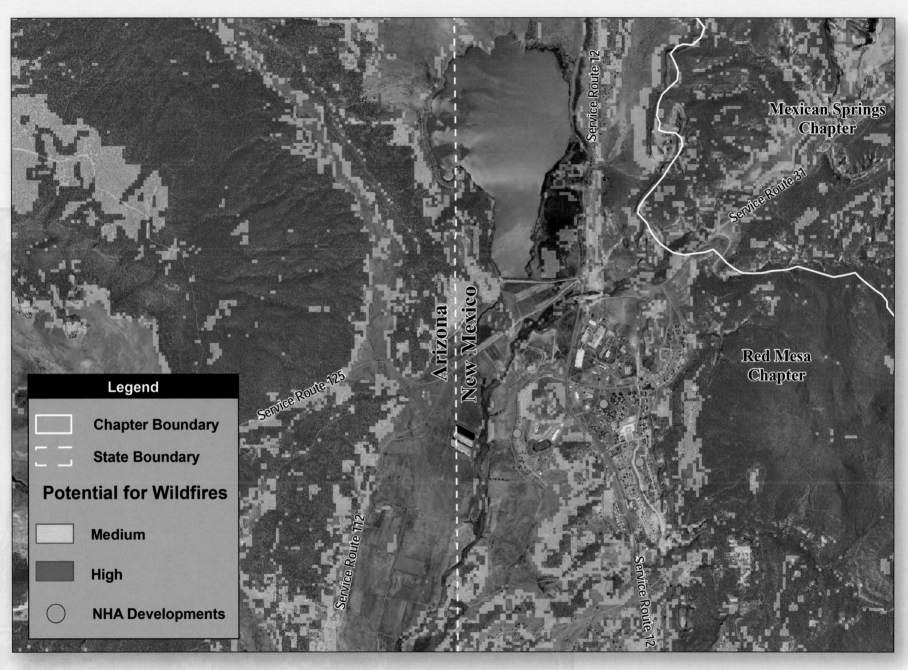

Legend

Chapter Boundary

State Boundary

Potential for Wildfires

Medium

High

 NHA Developments

Mexican Springs Chapter

Red Mesa Chapter

Arizona

New Mexico

Service Route 12

Service Route 31

Service Route 125

Service Route 112

Service Route 12

The NHA understands that the potential for wildfire exists in several Navajo developments. This map is used for planning safer communities that build homes in areas outside of high fire hazards. As a result of this wildfire hazard mitigation, NHA will lower insurance rates on its structures.

http://www.navajo-nsn.gov/

Red Mesa Chapter

CONCLUSION

GIS helps federal agencies analyze complex situations, visualize problems, and create geographic strategies and solutions. In the process, these agencies experience increased efficiency, reduced costs, and improved communication, both internally and externally. *Mapping the Nation: Supporting Decisions that Govern a People* features many examples of government realizing substantial returns on its GIS investment at a time when resources are limited and performance demands never higher.

Through web and mobile applications and cloud technology, the GIS platform is accessible and easy to use, enabling federal agencies to work faster and smarter. GIS helps them meet growing demands for accountability and transparency by engaging citizens through dynamic, interactive maps. The technology promotes sound decision making to manage the nation's natural resources, deliver health and human services, protect and respond to threats and hazards, and simplify day-to-day operations.

As vast stores of government data grow exponentially, GIS is there to clarify, analyze, and integrate the data within a geographic context. Thanks to advancing technology and a deepening commitment throughout the federal government to share data, a golden age of collaboration is dawning to benefit the nation.

CREDITS

INTRODUCTION

FAA AeroNav Products, created by and data from Federal Aviation Administration Air Traffic Organization, Mission Support Services, Aeronautical Navigation Products.

US DEPARTMENT OF AGRICULTURE

GIS Support Maps for National Level Exercise and National Radiation Exercise, created by US Department of Agriculture (USDA) Office of Homeland Security and Emergency Coordination (OHSEC) Emergency Programs Divisions; data from USDA, National Level Exercise (NLE) 2011, 2012 and National Radiation Exercise by OHSEC.

US National Arboretum Explorer, created by Agricultural Research Service, US National Arboretum and Blue Raster LLC; data from US National Arboretum.

Coconino National Forest Initial Attack Map Book, created by US Department of Agriculture (USDA) Forest Service, Coconino National Forest District; data from USDA Forest Service, CITRIX.

Golden-Winged Warbler Habitat Modeling Using Lidar Data, created by the US Department of Agriculture (USDA) Forest Service (Monongahela National Forest), West Virginia University, USGS (US Geological Survey) West Virginia Cooperative Fish and Wildlife Research Unit; data from 2007 USDA National Agriculture Imagery Program imagery, West Virginia University.

Best Management Opportunity Model—Restoring Chestnut to the Landscape, created by US Department of Agriculture (USDA) Forest Service; data from USDA Soil Data Mart.

Wallow Fire: Bringing Partners and Data Together, created by and data from US Department of Agriculture Forest Service Fire & Aviation.

Vegetation Height Santa Fe National Forest, created by US Department of Agriculture Forest Service; data from Forest GIS Group, Southwest Regional Office; Remote Sensing Group; 3DiWest.

Maah Daah Hey National Trail Map, created by and data from US Department of Agriculture Forest Service Region 1.

Forest Pest Conditions, created by and data from Forest Health Technology Enterprise Team.

Decision Support for Aerial Fire Retardant Avoidance, created by and data from US Department of Agriculture Forest Service Geospatial Service and Technology Center.

2010 Washington: Swakane, created by US Department of Agriculture (USDA) Forest Service, Remote Sensing Applications Center; data from Monitoring Trends in Burn Severity Project (MTBS), USDA Forest Service Remote Sensing Applications Center, and US Geological Survey Earth Resources Observation Systems (EROS) Data Center.

Fire Detections and Forest Service Lands: 2001 to 2011, created by and data from US Department of Agriculture Forest Service, Remote Sensing Applications Center.

Rapid Assessment of Vegetation Condition after Wildfire, created by and data from US Department of Agriculture Forest Service, Remote Sensing Applications Center.

Lidar Applications in the USDA Forest Service , created by and data from US Department of Agriculture Forest Service, Remote Sensing Applications Center.

Potential Areas Suitable for Forest Restoration Projects, created by US Department of Agriculture (USDA) Forest Service, Geospatial Service and Technology Center; data from USDA Forest Service.

The Interactive Visitor Map, created by US Department of Agriculture (USDA) Forest Service Geospatial Service and Technology Center, USDA Forest Service Office of Communication; data from Esri and USDA Forest Service.

USDA NASS Cropland Data Layer, created by and data from USDA National Agricultural Statistics Service.

US DEPARTMENT OF COMMERCE

Sea Level Rise and Coastal Flooding Impacts Viewer, created by and data from National Oceanic and Atmospheric Administration (NOAA) Coastal Services Center (CSC).

Offshore Wind Development Potential, created by National Oceanic and Atmospheric Administration (NOAA) Coastal Service Center, Bureau of Ocean Energy Management, Department of Defense, and DOE National Renewable Energy Laboratory; data from Esri Ocean Basemap, General Bathymetric Chart of the Oceans, NOAA, National Geographic, DeLorme, NAVTEQ, Geonames.org. Not to be used for navigation/safety at sea. This work is licensed under the Web Services and API Terms of Use. For information on specific bathymetric data sources used to compile this map, links where the data may be downloaded, as well as copyrights and use constraints related to that data, refer to the Ocean Basemap Contributors list.

Commercial Vessel Density, October 2009–2010, created by National Oceanic and Atmospheric Administration (NOAA) Coastal Service Center; data from NOAA's Office of Coast Survey, Esri Ocean Basemap, General Bathymetric Chart of the Oceans NOAA, National Geographic, and DeLorme. Not to be used for navigation/safety at sea. This work is licensed under the Web Services and API Terms of Use. For information on specific bathymetric data sources used to compile this map, links where the data may be downloaded, as well as copyrights and use constraints related to that data, refer to the Ocean Basemap Contributors list.

ENC Direct to GIS, created by National Oceanic and Atmospheric Administration (NOAA) Office of Coast Survey; data from NOAA Electronic Navigational Charts.

NOAA Gulf of Mexico Data Atlas, created by the National Coastal Data Development Center (NCDDC), a division of the National Oceanographic Data Center (NODC) and the National Environmental Satellites, Data, and Information services (NESDIS); data from NESDIS, NODC, National Marine Fisheries Service, National Ocean Service, US Army Corps of Engineers, US Geological Survey, and the Bureau of Ocean Energy Management, plus state agencies, nongovernmental organizations, academic institutions, and international partners.

Integrated Ocean and Coastal Mapping Sandy Coordination, created by National Oceanic and Atmospheric Administration (NOAA) National Ocean Service, Office of Coast Survey; data from NOAA, Federal Emergency Management Agency, New Jersey Department of Transportation Mapping Priorities, University of California, Santa Barbara.

Age 40 to 44 Net Migration Flows for Dane County, Wisconsin, created by US Census Bureau; data from US Census Bureau MAF/TIGER database, 2006–2010 5-year American Community Survey.

Percent Change in Population by Census Tract: 2000 to 2010, created by and data from US Census Bureau.

Ohio Congressional District 8 (Speaker John A. Boehner) , created by and data from US Census Bureau.

Congressional Districts of the 113th Congress of the United States (January 2013–2015), created by US Census Bureau; data from US Census Bureau MAF/TIGER data base, 2010 Census.

Small Area Income and Poverty Estimates, created by and data from US Census Bureau.

US DEPARTMENT OF DEFENSE

US Army, *The Battle of Wilson's Creek*, created by US Army Training Brain Operations Center; data from Esri; National Park Service; Knapp, George Edward. 1993. The Wilson's Creek Staff Ride and Battlefield Tour. Fort Leavenworth, KS; Combat Studies Institute.

Arctic Sea Ice 2007–2012, created by US National Ice Center; data from US National Ice Center (ice data) and Esri (basemap).

Post-Superstorm Sandy Elevation Differences and Beach Volume Change, created by US Army Corps of Engineers (USACE); data courtesy of Charlene Sylvester, USACE Mobile District at Joint Airborne Lidar Bathymetry Technical Center of Expertise.

Drought Status in the Fort Worth District, created by and data from US Army Corps of Engineers, Fort Worth District.

Extent of Flooding 1927 versus 2011, created by and data from US Army Corps of Engineers.

North Carolina National Guard Viewer, created by and data from North Carolina National Guard 13 Joint Operations Center.

Commander Navy Installations Command, *Navy Site Availability to Renewable Energy Resources*, created by Geographic Information Services, Inc.; data from Esri.

US DEPARTMENT OF HEALTH AND HUMAN SERVICES

Health Resources and Services Administration (HRSA), *Validating the Need for Additional Federally Funded Health Centers*, created by Blue Raster; data from UDS Mapper and HRSA.

Determining Safe Shellfish Growing Areas, created by Food and Drug Administration (FDA) Center for Food Safety and Applied Nutrition; data from FDA, Wastewater Effluent Dye Dilution Study, and Esri (basemap).

Global Spread of Disease Caused by International Travel, created by Centers for Disease Control and Prevention (CDC) Travelers Health Branch, Division of Global Migration and Quarantine; data from CDC.

Health Professional Shortage Areas—Primary Care, created by and data from Health Resources and Services Administration, Division of Policy and Shortage Designation, Data Warehouse.

Interactive Atlas of Heart Disease and Stroke, created by Centers for Disease Control and Prevention (CDC) Division for Heart Disease and Stroke Prevention; data from CDC.

Animated Historical Cancer Atlas, created by and data from National Cancer Institute.

US DEPARTMENT OF HOMELAND SECURITY

Fifth District SAR Cases Density Plot, created by and data from US Coast Guard Fifth District.

Winter Storm Nemo Forecast Impacts, created by US Customs and Border Protection, Office of Field Operations; data from Weather.com; US Customs and Border Protection, Office of Field Operations GIS; National Oceanic and Atmospheric Administration; Esri.

US Southwest Border, created by and data from US Border Patrol GIS Team (Washington DC).

US DEPARTMENT OF HOUSING AND URBAN DEVELOPMENT

Sandy Damage Estimates by Block Group, created by and data from US Department of Housing and Urban Development and Federal Emergency Management Agency.

US DEPARTMENT OF THE INTERIOR

Fort Ord National Monument Trail Map—June 2012, created by Fort Ord Project Office staff; data from US Bureau of Land Management, Fort Ord National Monument.

Oregon Central Coast Recreation Map, created by US Bureau of Land Management (BLM) Oregon State Office; data from BLM corporate.

San Juan Islands, created by US Bureau of Land Management (BLM) Oregon State Office; data from BLM corporate.

BLM Geoportal Server Map Viewer, created by US Bureau of Land Management (BLM); data from BLM Oregon State Office, Data, Applications, and Web Services.

Reservations with Significant Timberland Resources, created by US Bureau of Indian Affairs (BIA) Office of Trust Services, Geospatial Support; data from Intertribal Timber Council, 2003, BIA Office of Trust Services Division of Forestry and Wildland Fire Management, Branch of Forest Planning; US Geological Survey elevation data: 1993–2002; US Census; Esri.

The Hazard of Melting Permafrost for the Alaskan Natives, created by Southwestern Indian Polytechnic Institute; data from National Snow and Ice Data Center User Services, State of Alaska.

GLORIA Mapping Program, created by US Geological Survey (USGS); data from Coastal and Marine Geology Program of the USGS, GEBCO, National Oceanic and Atmospheric Administration, National Geographic, DeLorme, and Esri.

Known and Potential Streams at Morristown National Historical Park, created by National Park Service; geoanalysis and cartography courtesy of Roland Duhaime, Research Associate, Environmental Data Center at University of Rhode Island; lidar data provided by New Jersey Highlands Council.

US DEPARTMENT OF STATE

Ethiopia: Three Ways of Looking at HIV Distribution, created by US Department of State Humanitarian Information Unit; data from Central Statistical Agency (Ethiopia) and ICF International, 2012; Ethiopia Demographic and Health Survey, 2011; Addis Ababa, Ethiopia and Calverton, Maryland, USA; Ethiopian Health and Nutrition Research Institute Federal Ministry of Health; "HIV Related Estimates and Projections for Ethiopia, 2012," August, 2012.

Pakistan: Humanitarian Crises in 2011, created by and data from US Department of State Humanitarian Information Unit.

Syria: 2012 Population Displacement in Review, created by US Department of State Humanitarian Information Unit; data from US Department of State; US Agency for International Development; United Nations Office for the Coordination of Humanitarian Affairs; United Nations Refugee Agency; Internal Displacement Monitoring Centre; Syria Crisis Tracker.

World Humanitarian Day 2012: Aid Workers in Harm's Way, created by and data from US Department of State Humanitarian Information Unit.

Maritime Claim Lines and Limits near the Strait of Hormuz, created by and data from US Department of State Office of the Geographer.

Claimed 200 Nautical-Mile EEZ Limits in the Arctic Ocean, created by and data from US Department of State Office of the Geographer.

Syria: Halab (Aleppo) Governorate, created by and data from US Department of State Office of the Geographer.

January 18, 1974 Egyptian-Israeli Disengagement Agreement (Sinai I), created by and data from US Department of State Office of the Geographer.

Provinces and Districts of Afghanistan, February 2012, created by and data from US Department of State Office of the Geographer.

US DEPARTMENT OF TRANSPORTATION

Federal Lands Highway NPS Mapping Application, created by and data from US Department of Transportation Federal Highway Administration, Office of Federal Lands Highway.

Structurally Deficient Bridges, created by US Department of Transportation (USDOT), Research and Innovative Technology Administration; data from USDOT Bureau of Transportation Statistics, Federal Highway Administration's Office of Bridge Technology.

Draft Electronic Airport Layout Plan (eALP) for Valley International Airport (HRL), created by and data from US Department of Transportation Federal Aviation Administration.

ARTCC Low Sector Map, created by Federal Aviation Administration (FAA), Washington, DC; data from FAA Airspace and Procedures and Esri.

US DEPARTMENT OF VETERANS AFFAIRS

VA Influenza-Coded Outpatient Visits, 2011–2012 Influenza Season, created by and data from Veterans Health Administration, Office of Public Health.

VHA Health Care Supply and Demand for USVA Areas, created in collaboration with Esri by the Planning Systems Support Group within the Veterans Health Administration (VHA) to show location-allocation analysis in action to assist the Readjustment Counseling Services (RCS) with new Mobile Vet Center assignments; data from the VHA.

Health Care Safety Measures for the Veterans Health Administration, created by and data from the Veterans Health Administration.

VHA Health Care Service and Live Weather Events, created by Planning Systems Support Group within the Veterans Health Administration (VHA); data from the VHA, National Oceanic and Atmospheric Administration, and National Weather Service.

Using Location-Allocation Analysis to Improve Veterans' Access to Services, created by and data from Planning Systems Support Group within the Veterans Health Administration.

VHA Medicaid Expansion and Health Insurance Marketplace Web App, created by Veterans Health Administration (VHA) Planning Systems Support Group; data from DeLorme, Esri, National Geospatial-Intelligence Agency, National Oceanic and Atmospheric Administration, US Geological Survey.

VHA GIS Planning Portal, created by and data from the Veterans Health Administration (VHA) Planning Systems Support within the VHA.

INDEPENDENT AGENCIES

US General Services Administration Maps, created by US General Services Administration (GSA) Public Buildings Service; data from GSA Urban Development Program, Federal Emergency Management Agency, Esri, American Recovery and Reinvestment Act of 2009.

Tegucigalpa, Honduras Bus Routes, created by US Agency for International Development (USAID) Office of Transition Initiatives; data from OpenStreetMap, USAID.

Every Door Direct Mail Online Application, created by and data from US Postal Service (USPS).

USAID GeoCenter Maps, created by US Agency for International Development (USAID) GeoCenter working with other USAID entities; data from USAID.

Navajo Nation Maps, created by URS Corp.; data from Navajo Housing Authority, URS Corp., and US Army Corps of Engineers.

For more information on GIS in the federal government, visit esri.com/federal.